# The Queen and I

# The Queen and I

## Edward A. Weiss

*Illustrations by Calvin Diehl*

HARPER & ROW, PUBLISHERS
New York, Hagerstown, San Francisco, London

*Dedicated to Edna E. Erickson, an incomparable lady;*
*for her courage, her unbounding wit and good humor,*
*and most of all, her wisdom*

*Designed by Eve Callahan*

FIRST EDITION

LC 77-3779

ISBN 0-06-014578-1

78 79 80 81 82 10 9 8 7 6 5 4 3 2 1

# Contents

# Foreword

Is it possible to love an insect? The numberless books on honeybees declare in many ways that it is, but few say it with such sincerity and eloquence as this one. Sometimes it seems as though the tendency to fear and distrust is more easily roused in people than their capacity for love and appreciation, but this is usually because they do not understand. People think first of bees as menacing creatures, because they know too little of them to see and understand otherwise. The specter of "killer" bees approaching our shores from South America, which is largely the creation of journalists and film producers, preoccupies people, flourishing upon their fears and ignorance. Beekeepers themselves come to be classed in people's minds within the category of snake charmers and lion tamers.

In the pages that follow the reader will find a totally different and reassuring account of how one man, whose life had been caught up in the high-speed world of business, finally found fulfillment in his love for honeybees. He unabashedly speaks of them as his "girls," and in dozens of ways conveys the sense of their gentleness and beauty as well as his own love for the whole of creation. Nature is not a threatening jungle or battlefield, which we must approach with caution and against which we must arm ourselves. It is a heavenly place, which will fill our hearts with love and our minds with the sense of the glory of God. This Ed Weiss has seen, and the honeybee was his avenue to that vision. There is more to living than accumulating riches, more to success than large sales commissions, more to home than a place to rest from work, more to the human spirit than the drive to acquire, and more, much more, to honeybees than nasty little stings.

Nearly everyone has the urge to gardening and husbandry, the need to be close to the soil, to what is poetically called our mother, the earth. It is a need that is never quite extinguished by urban life and our accommodations to modern technology. Yet the demands of these are very strong, and it is easy to become trapped by them. And not only do the demands of modern life make it seem impractical to try returning to the simpler life of our ancestors; the conditions of contemporary agriculture also make it seem impossible. Our mobility, the need to be always coming and going, make it seem quite impossible to raise sheep or goats, and even the traditional poultry flock has been superseded by huge egg farms. Family farms have virtually disappeared, and as the barns collapse across the countryside, the farmhouses are taken over by urban commuters, this being the nearest they can come to the kind of life, close to the soil, that they inwardly yearn for.

The keeping of honeybees, however, still offers the same fulfillment of that need that it always has.

Almost anyone can own a hive or two, even if he has no place to put them except on his roof. They are no real menace to neighbors (though it might take a while for neighbors to realize this), and at a single stroke one is likely to find the whole direction of his life changed. It is no wonder that beekeeping becomes so absorbing, when one reflects that its product is the most delectable food known to mankind and that the tiny creatures who provide it are the most highly developed and inspiring representatives of the whole order of insects, or indeed, perhaps of all living things. Tending bees fulfills, quite completely, the need in all of us to raise things, to putter, to use our minds and our hands to practical advantage, and to dwell in wonder and fascination on the world of the honeybee that unfolds before us.

This is mostly a book for beginners, for those whose interest in bees has been stirred, but who have no idea how to become beekeepers. The author tells, in the finest detail, exactly what to do, and he has in mind readers who, like himself until about a dozen years ago, are utterly bewildered by apiculture.

Woven into his manual, however, is also the touching account of a deep love, and a tribute to the human spirit, in the author's description of the elderly and bedridden lady who initiated him into the beekeeping craft, and of the tender devotion to each other that evolved from this. No one will read his concluding description of this and ever forget it.

—RICHARD TAYLOR

# Introduction

The rich, life-giving soil has always beckoned to many of us. Especially at times when our lives seem to be rubber-band tight, churning with a steady sequence of stress-filled events that make one want to shout for relief, or perhaps just whisper a plaintive "Please let me slow down." Well, nature calls to everyone these days, in resistance to all the manufactured artificiality pumped into everything, including food, from preservatives to "pure" coloring.

*The search for a more natural way of life*

You may be one of those who want to shake loose from those frustrations in search of a more natural way of life. Perhaps your involvement in beekeeping is only one of many measures you are taking to achieve your goal of the good life.

If this is your objective, fine; but your motivation could also include a host of unknowns. Let's uncover a few so the urge for participation can overtake you, once they are exposed.

The honeybee produces many valuable products which can furnish any beekeeper with financial gain, if he desires to turn his hobby into a commercial venture, and an entire chapter has been devoted to them within this book. But I believe the other "profits" waiting for anyone who is involved in the husbandry of these wonderful insects are far more important.

*The benefits of fresh air and sunshine*

I refer to the complete relief from the turmoil and stress of our times one experiences while tending these faithful workers in the hive. Caring for the wonderful honey bee seems to generate and maintain good personal health. How many people avail themselves of the greatest gifts of nature: fresh air and sunshine? They are there, waiting for all of us, and beekeepers get their share as they attend their colony's requirements. In addition, the mild exercise derived from keeping bees seems just the right amount for the man or woman occupied by office work or some other sedentary occupation.

*Another product—a feeling that can't be sold*

Each time I work with the bees, it seems that I step into another world—a brief interlude from the tempest around us. I'm sure it happens to everyone who removes the cover of his hive, for it is impossible to think of anything else except these tiny creatures as you manipulate your way into their masses. The bees are highly communicative amongst themselves and in an uncanny way begin to communicate with you. I guess it sounds strange but it won't be long before you'll find yourself talking softly to them as you accomplish your inspections. I do, and I know they listen. Perhaps, in even a stranger way, I feel a spiritual kind of awareness envelop me and my bees when I am with them. You come so close to nature as you carefully avoid hurting even one wee bee. With great gentleness this care and consideration you have for them wells up inside of you to become a huge compassion for all of God's creations. And how they do respond! After several visits to any particular colony that I attend, I never consider that a single bee has in her mind to do me a hurt of any kind as I merely visit her world.

If I were alone with these deep inner emotions, I might start to wonder about myself, as perhaps you are doing now; but I can reel off the names of hundreds of people who feel just as I do. Even as I write this I remember one rather gruff man who came to me after a particularly severe winter to tell me about losing his colony. With tears in his eyes, he said slowly, "You know, I love my bees." Yes, love does creep in too. That is why I have seen wonderful things happen to many people, and inside me I shout for joy!

One cold winter's day, a man came to me to get the equipment to start up his first hive that following spring. He wore a long single-breasted overcoat buttoned clear to the top. It was a dark gray, matching the hat, a formal gray homburg. A long cigar jutted from his mouth, which scowled as he smoked and scowled again as he spoke. "Get out what I require to start up a hive." His overcoat was still fully buttoned and his hat stayed on. He was apparently an executive of some kind and accustomed to giving orders. Actually, he never relaxed on that visit any more than finally removing his hat and coat, which revealed a still more formal gray pin-striped suit, complete with vest and gold chain draped from left to right.

*The change bees made in one man*

I've described this gentleman in detail because this same man did become a beekeeper that spring and visited with me many times during that year. I watched the transformation of this man from the uptight, grim individual who first came to see me into an easy, smiling happy man, generally attired in casual work clothes. This chap, in his middle fifties, was talking about an early voluntary retirement the last time I saw him. If you are thinking that something else brought about this change, you're wrong. It was the bees. He had begun to deal with life, other than his own, and when he did, there was no room for his problems to occupy his mind; and that became a habit—a good habit.

If space and time permitted, I could recite many such examples of change wrought in people who reached out and were enveloped by their pets, the honey bees.

You might think it is sort of impossible for these plain old honey bees to do all this; well, it isn't. In fact, they do a great deal more, for as your interest grows in their welfare, your own fortunes will continue to expand. An awareness of many new crafts and interests will alert you to a new world.

*Bees will open up a new world of horticulture*

New awakenings in the world of horticulture will jolt you into seeing things you've never seen before. Did you know that the maple tree has flowers in the early spring? If you did, then you are already one of those special people. But anyway, they are there, sometimes all over the tree, sometimes way up on the top, early flowers with early nectar and pollen, ready for the bees to gather for their new brood, resting in their cradles, waiting to be born, to take their turn to gather nectar and pollen for those after them.

I never looked closely at a skunk cabbage before I became a beekeeper. When I learned that this lowly growth, inhabiting swampy areas, provides the honey bee with some of the earliest pollen, I took a good look. I learned that this plant, that blooms so early in the year, provides the bee with a warm room inside of the outer large petals while she gathers her pollen. Skunk cabbage has a good deal of starch in its roots and so provides extra heat within the plant. This suits our workers, who must venture forth to bring home the pollen sorely needed to feed the developing bees in the early coolness of spring.

Many other hobbies and interests creep in alongside of keeping bees. If you have never hammered a nail, you will find that your skills will develop here and you may find carpentry a new interest. Photography is a fine companion to beekeeping. The bees are just about the finest subject material a photographer could ever want. I know of some people who found that wax sculpting was their cup of tea. Their talents, long hidden, burst forth as the raw material was provided for them by the bees.

And then there is nutrition. Who can keep bees, have honey, and not become interested in nutrition? Hardly anyone can resist the inclination to find out all about honey.

A year or two after we became beekeepers and honey eaters, my wife and I saw our doctor for our annual checkups. It was usual for us to end up in his office, after all the tests and examinations were over, to hear what his investigations revealed. My wife, a long-time borderline anemic, always pre-empted him with a terse statement about it. "I know, Doctor, I need more iron in the diet and lots of $B_{12}$, right?" This time he looked at her strangely and said, "You know, Anita, I find no such condition exists any longer." She was astounded by his statement. We all

*Honey, a cure for one case of anemia*

wondered what had happened. Of course we learned that the tablespoon of dark honey over our cereal each morning for the previous year supplied what was needed to enrich her blood. We receive many hidden treasures from the bees and honey as the years pass; and you will too!

*This book's two goals are to share my joy and experiences, and provide a step-by-step guide for the beginner*

This book is my attempt to share with others the great satisfaction and pleasure I have experienced through my beekeeping. If only one more person in the world shares them with me, because of reading this book, I shall be happy. The book is also a step-by-step guide for the beginning beekeeper. It takes the new beekeeper through his entire first year of beekeeping. The book is basic. Absolutely nothing is assumed about the knowledge of the neophyte beekeeper.

It is for that reason that I make no attempt to be concise; instead my aim is to carry the new beekeeper from one step to the next without leaving an intermediate step out of the process. It is easy to say, "Light up your smoker and proceed to open the hive." I don't; instead I explain in thorough detail how to make a fire in your smoker so it stays lit throughout your visit with your bees. I certainly wish these steps and details had been provided for me when I began to work with bees. The book is arranged according to events as they occur chronologically. That which must be done first, I believe, must be explained and accomplished first to make ready for what follows. What to purchase, how to build it, and where to put it must be considered before the bees arrive. How the bees are bedded down when they first are put into the hive helps to determine whether the colony has a successful and productive summer. And to make it easier to find things quickly, in addition to the index I have also provided paragraph summaries in the margins.

As I have been a dealer in bee supplies to hundreds of people over the last few years, I have had to answer several hundred questions. The questions follow a pattern. None of us are so individual that we don't have pretty much the same questions as the next person. In the body of the book I have tried to answer those questions most frequently asked of me.

The scientists and the experts may not agree with everything that I advise (in fact, not all the experts agree with one another); but then, their points of view are different from mine, since they are not thinking in terms of helping the totally inexperienced individual who wants to become a beekeeper. Helping the beginner is exactly what I am trying to do.

# The Queen and I

# 1

## *How It Began*

I have been having an affair. You know the kind I mean: a love affair that carries with it all of the intrigue and romance one might imagine. It's a royal romance, and so my dear wife aids and supports the liaison. The queen and I will carry our affinity into the future. Her ethereal beauty has captured me. I am hopelessly hooked.

It seems like yesterday, though it is a time ago, that I looked for an address in Danbury, Connecticut. It was given in the Bee Supplies Catalog as 100 Elm Street. The name was E. E. Erickson. I had checked out the truth of the listing by phone. Someone—a secretary?—had answered and informed me that indeed this was where E. E. Erickson did business.

Elm Street is a pleasant, narrow residential lane, neatly representing its name with the trees fronting the old but well-ordered houses. One hundred Elm Street stands like a fortress, its square shape outlined in red brick.

The porch creaked its vulnerability, though, as I twisted the handle on the doorbell. I knew I was at the right place because of the faded sign out front on

Edna's house

an overgrown lawn, boasting "Bee Supplies."

The curtain behind the heavy glass door parted and a kindly-looking gray-haired lady inspected me. "Bee supplies?" I asked; and she nodded, opening the door. She directed me through a dark hallway into the parlor. My eyes swung around the room and took in the round oak table, the long sideboard with its lovely mirrors, an old-fashioned chandelier, a beautifully carved rocker, several old chairs, and a doorway with heavy drapes acting as the door. My eyes connected with a picture of an attractive young girl on the wall. The wood frame was handsome but as old as the costume that adorned the pretty girl in the picture.

"Would you like to become a beekeeper?" Her tiny voice was soft and gentle. My nod of agreement triggered this diminutive aged lady into motion. Wordlessly she hustled out a back door, returning with a box that had to weigh at least half as much as she did. She dropped the box, wheeled about, and disappeared through a different door into another room. I wondered if she had skates on, the way she moved; but I couldn't tell, for her gown was floor length and full. It was also up to her neck and long sleeved. The box she first brought in displayed a large letter A and a 1 alongside—that certainly is the beginning, I thought. She reappeared with her arms brimful of small boxes, wedging them all together with her chin. "I must go to the barn for the rest," and again she scooted out the back doorway. I looked quickly at the pile she had dropped on the table. I identified smoker, veil, hive tool, hat, feeder, and gloves with that brief look; and there she was again! This time she was transporting enough raw lumber to build the bow of a small boat. She wasn't even breathing hard. "My name is Doris, now listen attentively, I'm going to tell you all about this." I settled down, as she focused her blue eyes on me while cleaning her glasses. Doris put her glasses on and began.

"It divides up into three parts. The first part is the house for the bees; second, you require some tools to work with the bees; and third, items for your protection." She paused to rework a speck on her glasses. I was amazed at her head-on approach. For a moment I thought she might be oversimplifying. Then I

*A beginner's basic equipment*

considered her words. She was right. A hive is what is required to contain a colony of bees. A smoker and a hive tool are tools to work with; and I surely didn't want to be stung, so the gloves and the head veil.

Doris was steaming ahead. "This box contains a standard hive." I noted the first letter of the alphabet and the arabic 1 again. She continued, "The standard hive has almost everything to make a home for the bees. A bottom board, which is the hive base or stand. A hive body, which is this box open at top and bottom. It holds ten of these frames. You put sheets of wax foundation into these frames so the bees get a head start in making the comb. Then here is what will be an inner cover, as well as this outer cover, which is sheathed in metal on top.

"When you assemble all of this, you have the basic home for the bees, but you still need a front porch for them." She reached over into the pile of lumber and pulled out four pieces of wood which made up the "front porch" or hive stand.

"After you put your bees into the hive, they will go to work functioning as a community. A month or six weeks later, they will probably need a second story to allow for the increased size of the colony." The pile of lumber was disappearing as four more pieces of wood were mocked up into the shape of another box open at both ends. "So much for the bees' home. Here are the tools. The smoker and the hive tool are the two main implements." She stopped abruptly and walked through the curtains in the doorway into yet another room. I heard an exchange of low voices, but I could not distinguish any words.

I pulled on the pair of white leather gloves and

The "parlor"

thought about my hands being protected from bee stings. Then Doris came out of that mysterious room.

The curtains in the doorway were still swaying when Doris said, " 'She' wants to know why you want to become a beekeeper?" The reference to "She" puzzled me, but I responded to Doris: "I guess my primary reason is the need for pollination in our area. I understand that honey bees perform that function admirably." At that point Doris settled back and began a discourse which might have lasted a long time but for the interruption from the room behind the curtain.

"Young fellow, you need more than that reason to become a beekeeper." The voice was smooth, firm, and almost challenging. It was also definitely female. I threw a look at Doris and she whispered, "Answer her quickly!" Many thoughts rushed through my head—I knew the voice was that of a woman; was she E. E. Erickson? why the curtains? why was she hiding? why did she even question me? I was the customer and buying the materials; wasn't that enough? I moved toward the curtains and, instead of answering, asked, "What other reasons should I have, besides getting the honey the bees make plus the pollination?" There was silence! I felt a hand at my elbow; Doris was ushering me out of the parlor. In a front room she related a few quick facts: "Yes, that's Edna E. Erickson and she is the bee supply dealer. I help her in any way I can, and will go on doing so as long as I can. Edna is a great person; she'll help you get started in beekeeping, but you must never part the curtains or go into her room. Ask her all the questions you wish, but don't go into her room."

The message was strongly transmitted. I decided not to ask any further questions and just leave. As I carried the last pieces out to my car, I had to pass the curtains to "her" room. I paused a moment and said, "Goodbye, Edna." "Goodbye, yourself," came the retort. "Just remember, you must make the bees do what you want them to do and not let them get you jumping to their wishes," she admonished.

The thirty minutes or so it took to drive home gave me time to reflect upon the preceding hour's events. I knew I would visit that house many times in the future, but little did I realize the tremendous impact that Edna was to have on my days, months, and even years ahead.

# 2

## *Structures and Tools*

As I began to assemble the hive in my basement shop, my mind wandered off to that brick house on Elm Street. Perhaps Edna's question deserved a better answer. What did I really want out of beekeeping? Sure, I wanted our trees and flower gardens to prosper. The vegetables and fruits would do better too; and there would be honey, that beautiful liquid gold. But was that all? When the honey is harvested, the cappings on the combs are removed, then melted down into cakes of beeswax. Candles flickered through my mind. Perhaps I might share in the bees' harvest of pollen. After all, they feed it to their larvae as the protein source. I began to think of all the other bee-produced gems: royal jelly, propolis, queen bees. I learned later that all this was insignificant when compared to the considerable magnitude of the never-ending intangibles.

The opened box containing the "standard hive" proved to have a neat booklet of directions, which I followed. The hive body went together easily. I matched the jigsaw corners, saw to it that the hand holds all faced up, and drove nails into the holes

*Building the basic structure*

drilled to receive them. It couldn't have been easier! The bottom board (which the hive body sits on) is really two long rails, each with a neatly grooved channel to receive the floor boards. A scribed line on each channel told me where to place the nails. Seemed easy too! Then came the inner cover. I formed the "picture frame" for the inner cover by assembling the four sides and then dropped in the flat pieces of wood where the picture would go. When nailed together, I had an inner cover. The outer cover was much the same, plus a galvanized steel tray which fit over the top of it. It was going better than I expected.

*Assembling the frames*

Then I went back to the box to see what was left. Here was the kicker! A collection of odd-shaped pieces of wood, and none seemed to match except the long rectangular sticks. According to the directions, the collection of wood when put together would make the ten frames which fit into the hive body. The pictures in the directions solved everything, and the frames were assembled—but they weren't easy! I bent so many nails that I ran out of them. Later, as I explain on page 14, I learned how to do it better. With the frames assembled, I was finished except for inserting the wax foundation into the frames, and assembling the hive stand on which the bottom board sits.

*Assembling the hive stand's front board*

I tackled the hive stand, assembling the back and sides first. Easy. Then the front was attempted ... not easy, but eventually I learned. The hive stand's front board, the one that slants for the bees to land on, always provides a nailing problem. It's easy to solve if you assemble the three other pieces first. Then put the back piece up against a wall (or stop) so that the whole thing can't go anywhere. Next, put the landing board against the two runners sticking out. Nail one side first. Put one nail into the center; then the one above it. Finally comes the tough one, the bottom nail. It's the angle that will throw you. You cannot follow the angles of the first and second nails, because you are so low that you'd end up nailing the stand to whatever it's sitting on. Start the nail by tapping it gently, while holding it on a line parallel to the ground or work bench. Then drive it home. By virtue of assembling the hive stand on a flat surface, you'll have no wobbles later.

Now for the wax foundation, which comes in rectangular sheets that fit into the frames. Of all the various forms of foundation available, the easiest and best for the beginner to use is crimp-wired foundation. This type of foundation has vertical wires running through it, curved into hooks at one end. The instructions called for breaking out a strip of wood called a "wedge bar" that is located on the under side of the top bar of the frames and then, when the wax foundation is inserted into its proper slot, nailing the wedge bar back against the top edge of the wax to hold it in place. Great!—if you have four hands. Hold the wax, hold the wedge bar tight, hold the nail—and then hammer it in. The first time I finished the job with my left thumbnail ready to depart, and four sheets of wax with round holes from a hammer blow. As you can guess, a hole isn't desirable.

*Adding the wax foundation*

Take heart, because there really is a better and easier way. First of all, purchase a brad driver. The bee-supplies dealer has them. It is a handy tool. Pick up some ¾-inch—18-gauge—brads. Even at today's inflated prices, they are a bargain. Fifty-five cents will buy a whole box. Enough to probably fill a hundred frames with foundation.

Positioning the sheet of wax foundation into the frame is frequently a puzzling maneuver for the neophyte. There are two ways of proceeding, and it seems to me that most folks choose the difficult way because it appears easier until you actually try it. The best way in my experience is to insert the bottom of the wax foundation (where the wires do not protrude) between the two bottom bars by holding the frame in your left hand with the top bar up and the empty wedge bar space facing you. With your right hand grasp the sheet of wax foundation by the top with the vertical wires of the crimped foundation facing into your hand. Now you are able to drop the bottom of the wax sheet between the two bottom bars. As you raise your right hand the crimped wires will fall into place against the empty wedge space of the top bar. When the frame is turned upside down, with the top bar on the table, you can adjust the lateral position of the wax foundation right or left to make it equidistant from the side bars before nailing the wedge bar on top of the crimped wires.

If you are working on a regular work bench, nail a strip of scrap wood to the bench to act as a stop to push the top bar of the frame against. The frame will then be upside down against the stop, with the wedge bar facing you. Now that you have the sheet of foundation in place, position the wedge bar against the foundation, sandwiching the foundation's crimped wires between the wedge and the top bar. Load the brad driver. Now, with your left hand pushing against the wedge bar and the brad driver

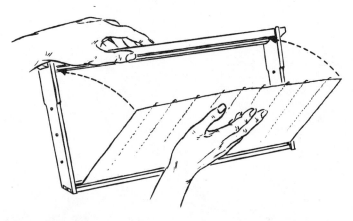

Insert foundation between bars of frame and raise till hooks rest against top bar

in your right hand (if you are not left-handed), drive the brad into the center of the wedge bar. Move to one end of the bar and repeat the process; then to the other end of the bar. Finally put in a brad halfway between the center and each end. The result is five good fastenings and no punctured foundation . . . and only two hands did the job. By the way, it's easy to improvise some kind of stop wherever you do your work—except on the dining-room table. A piece of wood and a couple of clamps will do the job anywhere. No clamps? Put your table against the wall and use your wall as the stop.

Now, let's stack it up and see how the bees' new house will look. Place the bottom board on the stand. Does it wobble a little? Don't worry, it will settle down. Now comes the hive body. Its bottom edge on three sides matches the bottom boards' three ledges, forming an entrance slot in the front. If it rocks on the bottom board, look for the high spot and sand it

*Putting it all together*

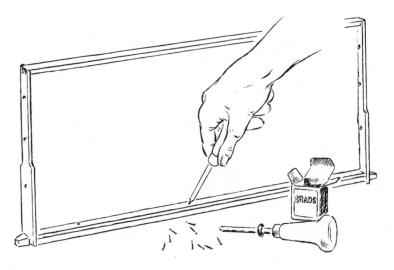

Wedge bar fastening the easy way—use a brad driver

down with some coarse sandpaper. It must not rock. Ten frames into the hive body and on goes the inner cover, flat side down, ledge side up, half-moon hole open. Finally the outer cover is placed on top. Stand behind the hive and push the outer cover forward; that opens the half-moon ventilation hole in front. Pull the cover back; that closes it. Leave it open except on certain occasions, which we'll talk about later. Again standing behind the hive, adjust the outer cover so the spaces between the edge and the hive body are about the same on both sides.

When I had finished, I still had the second hive body to build and decided to do it then rather than later. Good thing I did too, since I had the opportunity to paint everything at once rather than piecemeal.

*A carpenter's square*  Besides the brad driver, invest in a carpenter's square. It will solve the rocking problems and the problem of excessive gaps or spacing, which can turn out to be quite an annoyance when your colony of bees and its honey supplies are being attacked by robber bees. Too wide a space between hive bodies also allows the cold winter wind access to your many thousands of workers struggling through the long winter months.

Here are a few things to watch out for—and how to avoid them.

To keep the hive body square while assembling it, put four nails, one in each section near its center, before putting all the nails into each side. If you throw an arm over the hive body you can hold the box joints tightly together for the first nail. Then, using your square, put the second nail next to the first. Doing it this way, the box (hive body) will end up square. If the top or bottom edge of one side is higher than that of the side next to it, now is the time to sand or plane it off. That, plus the squaring of the hive body, eliminates the rocking, which eliminates the robbing problem; which eliminates the cold-winter-winds problem; which means you've helped the colony's chance to survive.

*Keeping the hive body square*

When a real house is constructed for us humans, it has outside walls and a roof; then, by adding partitions, we form rooms to perform various functions: sleeping, eating, cooking, and all the other things. The frames which are put into the hive body are the rooms that the bees use to perform their multitude of functions. Each frame has a single sheet of foundation in it. The bees make wax and develop approximately thirty-five hundred hexagonal cells on each side of the foundation some three-eighths inch deep.

These cells contain, at various times, eggs, larvae, pupae, pollen, nectar, water, and honey. The weight carried downward by all these within the comb (the expanded sheet of wax foundation) is enormous. If you secure the top of the foundation within the frame properly, the bees will do the rest.

*Tips for assembling frames that will last*

The frame itself must withstand many rigorous strains, so construct it properly without short cuts. If you bend a nail and hammer it flat to get rid of it, you've cheated, and sometime or other that joint will open up; the nail must use its holding power over its entire designed length. Here is a way to avoid bending nails and assemble a frame that will last and last.

Pick up an end bar and place your left thumb on the beveled edge while you take the long top bar and fit it into the groove provided. Pick up another end bar and again place your left thumb on the bevel. Put the other end of the top bar into it. Don't nail anything. Put the frame down with the end bars sticking up in the air. (The frame will be resting on the top bar.) Insert the two bottom bars into the slots. Now make the bars flush with the outside of one end bar. Drive a nail into each. These nails go straight and easy. Now flush the bottom bars at the other end bar and nail them.

Turn the frame over so you are looking down on the top bar. The next four nails are the tough ones. Two nails must go through the top bar into the end bar. They are side by side and no pilot holes are drilled. Hold a nail between your thumb and index finger in place and, unlike a carpenter, tap the nail gently and repeatedly till the head of it is almost at your fingers. Frame nails are very thin and bend easily when a great length is exposed, but now only a small bit is exposed, and with a blow or two the nail will go directly home. These nails are rosin-coated and have tremendous holding power, even though only 1¼ inches long.

The left thumb caper provides you with frames all positively interchangeable in any of your hive bodies. Always use the left thumb against the beveled edge of the end bar and you will never be bothered by having two flat sides of adjacent frames glued together by the bees. It is practically impossible to pry glued-together frames apart without breaking the wood.

There is just one more detail. I don't know why the manufacturers don't do it. Make up four small squares of wood from the thin end of a shingle or some similar piece of wood. They should measure roughly ¾ inch square. Secure them with a waterproof glue to the four corners of the flat side of the inner cover. This is the non-ledge side. The four corner pieces should elevate the inner cover about 1/16 inch from the hive body. This will permit proper ventilation both winter and summer. Though there are many ways to effect ventilation within the hive, it appears to me from all manner of experiments that this works best. The need for ventilation in the winter we'll talk about later when we discuss wintering, but the squares are effective in keeping the colony dry when moisture is present. Summertime heat can be a cause of swarming, so the procedure helps at that time too.

*Four small squares of wood for ventilation*

Now that it is assembled, we are ready to paint the hive so it stands up under hot sun, moisture, and the truly tough rigors of winter. Water-based latex paints

*Painting*

offer several advantages and we use them. The color is up to you as long as you keep it light. A medium green is okay but don't use a dark green or brown or black. Dark colors absorb too much heat come summer, and the bees use up too much energy cooling the hive. They are air-conditioning experts. Many pounds of water are brought into the hive, placed in the cells and evaporated by relays of fanning bees. The water carried into the hive could have been nectar instead.

Anyway, put two good coats of paint on everything you can see, when you have once stacked up the hive. The interior of the hive need not be painted. In fact, the bees prefer you don't paint it. They like the feel of wood. Their tiny claws like the roughness of even a smooth wood surface. The one thing you might paint that you don't see is the inside of the bottom board. The bee mistress, in the guise of my dear wife, and I have had a running controversy about this for years. I think I'm beginning to agree with her. The bottom boards do take tremendous use and abuse; so paint them if you choose, but not the rest of the inside of the hive.

*Bee equipment terminology* The hive body has many names. At various times you might hear it called a "deep super" or "high body." Some think of it as a "story" or "floor," as in a building or an apartment house. It is so much nicer to call that first hive body a "brood chamber." It really is just that, a chamber where your lovely queen and her multitudes concentrate on raising thousands of baby bees to replace themselves as their lives ebb and flow through the seasons. The next hive body, which is not put on until after five or six weeks, is called a "food chamber." It is here that later in the summer the colony begins preparation for winter. Almost every frame in the food chamber is filled with honey and capped over to keep the moisture out. If all goes well for your tiny charges, this reservoir of honey that you leave them with will fuel their heat-generating machines through the cold of winter.

Once the food chamber shows promise of filling (60 to 70 percent filled), you will add a queen excluder and place a shallow body above that. The inner and outer covers move upward regardless of the hive size. There is much debate about the use of queen excluders. We will discuss the pros and cons later when I tell you the secrets of manipulation. The shallow body, or shallow super, that is placed on top of the food chamber will contain the surplus honey—and that is yours. More than one will be added in a good season, to build your skyscraper. It is possible to use deep supers, but I consider the use of shallow supers advisable for us because of two things, and they are related. Backs and weight. A deep hive body filled with honey, comb, bees, and all can weigh close to one hundred pounds; a shallow super will weigh around fifty to sixty pounds—certainly heavy enough for us two-legged animals.

*Queen excluders and surplus supers*

The use of shallows also helps one become less clumsy. Less clumsy means fewer or even no stings. Your bees will sting you only as a last resort. A worker bee (female) is the only one equipped with the stinging mechanism. The drones (males) do not have stingers and the queen will only use her stinger on another queen. The queen's stinger is smooth and therefore she withdraws it and lives on. Not so with the worker bee, whose stinger is barbed and when

*Bee stings: Why and when*

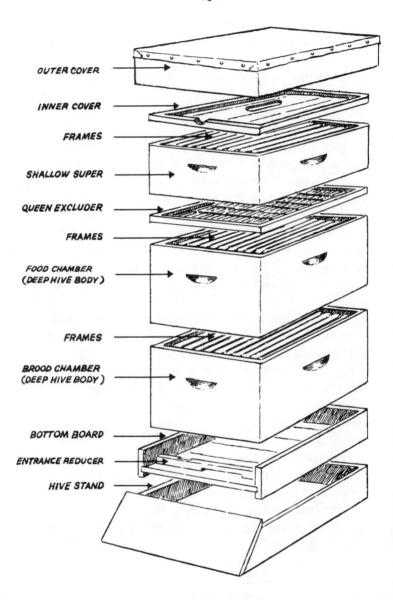

OUTER COVER

INNER COVER

FRAMES

SHALLOW SUPER

QUEEN EXCLUDER

FRAMES

FOOD CHAMBER
(DEEP HIVE BODY)

FRAMES

BROOD CHAMBER
(DEEP HIVE BODY)

BOTTOM BOARD

ENTRANCE REDUCER

HIVE STAND

The bees' abode—with surplus shallow super

used stays in the victim. When our hard-working lady flies away, a section of her innards stays with the stinger, and so she soon dies. Her life is never given without cause. She will not sting indiscriminately. There must be a reason. If her home is being invaded, if she is squeezed, alarmed, or frightened, there is cause to defend the colony. The honey bee simply does not sting for sadistic pleasure. She, unlike the wasps, can only sting once. The venom sac is connected to the stinger, and it is this part of the body that is torn out. It remains on the surface of the stung person's skin, pulsing from a set of involuntary muscles which continue to squeeze venom through the stinger. Scrape it off with a fingernail or knife. Even a packet of matches will separate the sac from the stinger, thereby minimizing the reaction.

*The smoker and how to use it*

Well, the big pile of boxes and lumber I had purchased to start my hobby of beekeeping had practically disappeared. The gloves, head veil, hat, and hive tool had not yet been needed—and there was that fascinating little machine, the smoker. It really intrigued me, that shiny tin can with the bellows attached to belch fire and smoke. Only later as a seasoned beekeeper did I learn how comfortable a companion Mr. Smoker becomes. If someone had only told me how to use it, I guess I might never have had half the stings.

Fire building has never been my strong point. I remember how, many, many years ago as a Boy Scout, I earned the nickname of "Five-Match." We were on an overnight hike and I was engaged in taking my test for fire building and cooking. We were allowed two matches. I suppose you can guess the rest. I flunked four times and passed on the fifth

Smoker and hive tool

attempt. Now I am a Fire Builder De Luxe. I learned how because nothing will exasperate the beekeeper more than a smoker that suddenly goes out—with the hive wide open, halfway inspected, and black with bees.

I have found that to start a fire nothing burns better than a piece of paper. I'm partial to paper towels. Two pieces are loosely stuffed into the smoker. Then come the "toothpicks." These are dry twigs from branches of fallen trees, which are slender and crisp like a toothpick. Make up a small bundle of these pieces about as long as your smoker can. Twenty or thirty pieces will do. Then collect ten to fifteen index-finger-sized pieces of wood. Finally find three or four dry pieces about one-half the diameter of your wrist. Place, don't throw the fine

Smoker fuel "by the numbers"

twigs on top of the paper. Now, with one match, light the paper. I usually light the match, get it burning a quarter of the way on itself, and then, using my bombsight, drop it into the smoker to reach the paper.

As soon as you see the paper burning, start working the bellows. Now here is the trick: don't work it violently; use rapid short squeezes. Keep this up until you hear a crackling sound. That means the wood is beginning to burn. Paper burns softly, almost without sound, but not so wood. Keep up the short pumping motion until you can see the wood is burning. Lengthen the pumping stroke of the bellows for five or six strokes and then stop. The flames should continue to grow. When you are certain the wood is burning and can take it, add the index fingers, one at a time. Give the bellows two or three squeezes each time you add a piece of wood. When these pieces are used up, add the three or four pieces of heavy stuff; but again, one at a time. Remember to squeeze the bellows a couple of times whenever you introduce a stranger.

If your wood selection was proper, you'll be looking for space to put the pieces in. Don't worry, you are on the right track. Jam the pieces in every two or three minutes, each time giving the bellows a few pumps. Don't close the cover because you are not finished. Let it burn this way (jammed tight with wood) for about ten minutes. Then, with the cover still open, pick up your smoker and rap it down against the ground two or three times. This brings all the coals down and pulls together the wood yet unburned.

Now you are ready to add the real fuel. Here you have several choices. My favorite is coarse sawdust. It is easy to get and keep on hand. I always carry a box with me in the car and whenever I see state workers cutting down a big tree (or anyone else) I stop and fill up with sawdust. If you have access to a carpenter shop or sawmill you can always get a supply. My second choice is baling twine. Farmers receive hay baled with twine. They'll give it to you gladly. I also like the husks of our shagbark hickory nuts. In the fall of the year, the squirrels help me gather a supply. They crack the nuts and I collect the husks. Then my bees are hickory smoked. Punky wood can also be used, but it doesn't last as long as the others. You can burn burlap too; but I found it expensive and too quick in the burning. It is good for a short visit to a hive—perhaps ten to fifteen minutes—but for a long-term smoker, use the sawdust. Regardless of which fuel, add clear to the top of the can, close the hood, pump a few times, and you have a smoker that will burn from eight in the morning until five in the afternoon with clouds of heavy white smoke. You can see that I start to prepare my smoker about a half hour before I know I'm ready to use it.

Honey bees are easily handled when a smoker is used properly. Further along, when we open a live colony, I'll explain how to use this tool. Many opinions are expressed as to why the smoke makes the bees compliant and easy to control. It seems reasonable to believe that the bees' instinctive response to smoke is associated with forest fires. After all, the natural home of the honey bee is in a good-sized tree. They build combs in the center of the trunk of the tree. They consider themselves well protected from everything; and so they are, except from a forest fire. When smoke is detected by the bees in their natural tree home, they fan violently, trying to keep the interior cool. They also run for their treasure under the mattress. All hands push their heads into the cells and suck up as much honey as their honey stomachs will hold. This is their treasure to be transported with them, if need be, to a new home. The honey stomach is equivalent to their traveling bag or valise. The bees have two stomachs. The other is a digestive stomach.

So, when you puff a little smoke at your bees, they then busily engage in filling up, in case they have to leave. They pay you little heed. All heads are in the cells and all little behinds are stuck up into the air, instead of in you. The honey is not lost, since it is returned to the comb as soon as the crisis is over.

# 3

## The Amazing Bee

I thought a great deal about Edna Erickson during the next few days, especially about her words of wisdom, "You must make the bees do what you want them to do and not let them get you jumping to their wishes." I realized I would have to visit Edna again. Perhaps she would clarify her dictum on how to control the bees. Perhaps there were secrets.

Speculating on how I would get past Doris to talk with Mrs. Erickson, I rang the doorbell. Doris peeked from behind the curtains on the door. She apparently recognized me, for the door swung open.

I followed her to the parlor and explained my need for frame nails since I had bent so many. She chuckled and pulled out a drawer full of various kinds of hardware that included nails. As she handed me a packet of nails, I told her that I wanted to talk with Mrs. Erickson. Instantly from the other room Edna loudly proclaimed, "It is Miss Erickson, young man." Quite humbly, I stuttered my embarrassed answer of compliance. Then through the curtains came a quiet "How can I help you?" I explained I was concerned with what was ahead, how to develop a technique for handling the bees. I told her

I had the feeling she knew how. There were a few moments of silence before she began.

"Pull a chair over to the curtains, and from now on please use my first name. I think we will have a good many talks ahead of us." I was pleased we had established a beginning and sat down in the chair provided by Doris, who promptly disappeared. How could I have known that this was the beginning of a rare relationship? This courageous lady would change the direction of my life one hundred and eighty degrees.

"Your interest in technique is best served by understanding these little creatures," she began. "This understanding begins the moment you realize that you and the bees have identical objectives: honey and pollination. The bees gather nectar and make honey for survival. A worker bee visits hundreds of flowers each day in her quest for nectar. Pollen grains stick to the hairs on her body and are carried from flower to flower. Of course, bees also collect pollen to feed the larvae in the hive. Pollen is their source of protein, just as nectar is the carbohydrate portion of their diet. When they mix the two together, it is called 'bee bread.'" She stopped and offered me a honey candy, poking her hand through the curtains. "Have one; they're good for you." Edna sounded so positive about everything, but not really bossy. I knew from my reading that bees pack pollen into the cells for use during the current year and continue even after present demands are met. The excess is stored over the winter to be available when late in the winter the queen begins laying eggs again. Naturally there are no blooms from which to harvest pollen in late winter, so during the summer the bees anticipate for the following spring.

*Under-standing bees*

I took the candy, gave her hand a squeeze, and asked her to continue. She said, "Why not team up with the bees since you are both seeking the same goal? Understanding the bees will help you learn a *concept*. Learn to recognize each stimulus that makes the bees respond in a certain way. For example—if you went to a hive and gave it a kick, the bees would probably rush out to defend their home against the invader, and you'd be stung in the process. Human beings think they're clever. What follows your hunger? Eating! Which is your instinctive response to the stimulus of hunger. Touch something hot and you pull your hand away without thinking. Stimulus-response again. These stimuli are many and they are, too, for the bees. Once you learn what the bees react to, you'll begin to *think like a bee!*" Edna caught her breath and then said, "Ed, you probably think I'm odd, saying that." Her remark took me by surprise, because I had just thought: How does one think like a bee?

"It is a new idea for me," I answered, "but no, I don't think you're odd. You know something the rest of us don't. I need a little time to absorb your interpretations."

Doris suddenly nudged me and I knew Edna's time with me had ended. As I prepared to leave, she told me to learn as much as I could about the "players." I guessed she meant the worker bees and drones; and, of course, the queen. I thanked them both and left.

I always try to simplify any new venture. I believe we all look for the basic requirements to make each endeavor a success. I had come to realize that in

Total concept: beekeeping

beekeeping, as in many other areas, nature plays the most important role. Bees need sunlight, wind, and rain for the plants and trees to grow, to nurture the flowers that spew forth the nectar that the bees convert to honey and the pollen that feeds the new populations of bees. Certainly we need to perfect techniques for handling the bees properly, to give them our assistance; but without nature's help we don't stand a chance.

*The queen*    We need a good queen bee to lay the eggs that hatch into the bees who populate the colony. The queen is paramount. She is supported—literally fed—by the workers who make up the bulk of the population. Finally, there are the drones. Now that seems to round it out. All the ingredients required for beekeeping, plus you and love.

*The drone*    The drone is a fascinating fellow, often maligned by beekeepers as "rather unnecessary," "a big eater," and "lazy." Well, all of this may be true, but without him there would be no bees at all. It is he who mates with the queen "on high." It is rather interesting to note that the poor guy has one good fling. The virgin queen leaves the hive to be mated and flies to some two to three hundred feet; the first drone who catches her joins her body in the mating act and then literally explodes. When he falls away from her, a portion of his reproductive system stays with hers. He dies following this event. A strong flyer, and by the size of his wings and wing muscles he shows it. He is even capable of carrying a queen along while mating. The drone's eyes are extremely well developed and for the most obvious reason. He must be able to spot a queen in the air at mating time. This wonderful fellow does not have a stinger; but, oh!

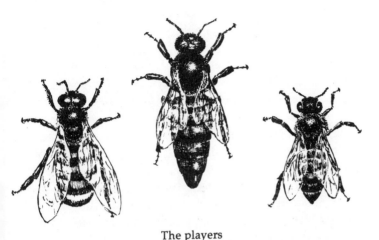

The players

does he make a great deal of noise when he flies by! He is larger than the workers, with a squat rear end, large eyes and wings. Beginners frequently mistake him for a queen; but the queen has a tapered abdomen, which is unlike his squat end.

The drone serves the beekeeper in yet another way. The workers tolerate his presence as long as food stores are good and plentiful. Should they find food scarce, the drones are driven from the hive. This does happen once in a while during the season; but usually it is at the very end of the nectar-producing part of the summer. So when you see the drones being pushed from their home, you'll know the season is over. The longer the drones are permitted to be around, the more prosperous the colony.

Whenever I see a queen, while inspecting a colony, I seem to see a heart in a body. Perhaps it is because I feel that if a colony were compared to a

*The worker and the queen*

human body she would surely be the heart and the workers would be the other organs and cells. Most of us know about the queen's main function, that of laying perhaps as many as two thousand eggs in a single day. Her long and slender abdomen seems to belie even the possibility of laying that many eggs. Where could she store so many? Why, they probably weigh almost as much as she does. She is very special and her worker bees know it, for they feed her, clean her, and attend her in every way. The constant feeding of a highly nutritious food known as "royal jelly" permits her system to manufacture the required number of eggs. Her capability is diminished if the quantity of food supplied to her is lessened or ceases altogether.

Now stop a moment and digest that. This means that the worker bees really control the output of the queen. After all, the worker bees know better how well the food (nectar and pollen) supply is coming in. If there is not enough food, why create more mouths to feed? In this way the population is regulated, by availability. The opposite is also true. If it is a time of plenty and the nectar is squirting, then the hive needs more foraging bees to get that nectar. So the bees press the queen to lay more eggs by gorging her with royal jelly. A nectar flow is a stimulant to a queen.

*The influence of the seasons* As a result of these checks and balances, the population of a colony will increase, beginning with the end of winter, and peak for the fruit-blossom period in May. This is the reason that May is generally known as the swarming time. If your girls are neglected by you and not given enough space by the addition of hive bodies in time, they will split the colony. About half the bees will leave with the old queen. The parent colony will survive, for the bees leave behind several queen cells from which a queen will be born. Your chances of getting a surplus of honey from that colony are seriously jeopardized. Give the bees the necessary room at the right time and they'll stick with you and be a strong colony.

The queen is not just a machine to lay eggs. She is responsible for evoking a morale response. Take a queen out of a hive and the bees will be quick to know it. Sometimes you can hear a low wail, mourning the fact that they are queenless. The queen secretes many different kinds of odors, which are called "pheromones." These pheromones are very powerful stimuli. One of these, called "queen substance," tells everyone in the colony that she is present. Take her away and you take away the scent. The bees become morose, dull, and listless. They even slow down the activity of foraging; fewer and fewer bees appear to be going in and out of the hive. The feverish excitement that goes along with a queenright colony does not exist. Most bee activities are aroused into action by one of the many pheromones. The worker bees are capable of producing message-sending pheromones also.

*Phero-mones—the powerful stimuli*

Colony inspection time is a happy time for me. My bee yard is a place where I feel a sense of peace. It is usually quiet, except for the sound of the tiny creatures at work. Their hum, which I must admit increases its intensity when summer nectar flow is high, soothes me. My communion with the bees reaches the heights when I'm holding a frame just drawn from the hive.

*Inspecting your colony*

My eyes follow the individual bees as they move

through the frame performing their various functions. I cannot feel other than wonderment when I see a single bee perform three or four tasks in just about as many minutes. I feel some sadness too when I realize that even the newborn chewing her way out of a capped cell will surely be dead in six weeks, perhaps sooner. Her work is hard, from birth to death. They are a no-nonsense society.

*A bee's life span and duties*

A honeybee is twenty-one days in the making, all of three weeks; and then she has just twice that time to live an active life. It seems hardly enough time to execute the many functions that one single little insect must perform. Take the first three days of her *Cell cleaner* life: totally obligated to clean and polish recently vacated cell walls so the queen can deposit more eggs. There are about eight generations of bees raised each season, and the cells must be used each time.

*Nurse and dietician*

When our cell cleaner reaches her fourth day of life she becomes a nurse and dietician. During the next week of her life she tends the larvae, feeding them the honey and pollen mixed with enzymes produced by her own glands. The first part of the week she can only feed the older larvae, since the food she makes is a little coarser than the food fed to younger larvae. Her enzyme-making gland will develop fully in a couple of more days; then she is permitted the feeding of younger larvae. This gland is in the forepart of her head and is called the "hypopharyngeal gland." It produces an enzyme called "invertase," plus other digestive enzymes. Invertase changes sucrose into glucose and fructose.

*Attending the queen*

Now that she is a week old, she is allowed to make and feed royal jelly to the queen. This duty will occupy the next four or five days. She will attend the queen whenever her highness passes near.

After this session with royalty, our aging worker (nearly two weeks old) must now do the more arduous tasks of general housekeeping. Some of these jobs include cleaning, guarding, cooling and heating, evaporating nectar, making wax, sealing cracks with propolis (bee glue), building comb, capping honey, and more!

*House-keeping*

On the eighteenth day, and for two more, compulsory guard duty is her only job. This occupation entails great risk. If the time is during a dearth of nectar, then robbing is prevalent and that means a fight to the death to keep invaders out.

*Guard duty*

Now, with life half gone, our heroine has earned the right finally to leave the hive. Oh, she might have had a play flight or two, but now she goes foraging afar. She might collect pollen for the brood or even nectar. Not both at the same time. Or she might be detailed to bring back water for cooling down the hive, if the weather is hot and dry. So away she goes each day to collect and return, collect and return, until some three weeks later an event occurs. It might be a sudden rainstorm, or perhaps a bird. The dangers are many. Perhaps none of these hazards will befall this tiny bee, but instead her lacy wings become shredded and worn and finally cease to function. Her cycle is completed, the six weeks end.

*Foraging*

The worker bee is totally adaptable, ready for change as required. She truly rises to the need of the moment. As architects bees build all structures required, with bridges and girders and more. Then they play the role of dieticians, making royal jelly and other foods for their larvae and queen. Then,

Architects,
dieticians,
air-condi-
tioning
experts,
defensive
specialists,
navigators,
honey
makers

too, they are marvels when it comes to heating and cooling. They are probably the original air-conditioning experts. All summer they bring in water to cool the hive through evaporation. Military experts, detailing an entire defense system, and navigators— why, they even use the sun to take azimuth bearings to locate a source of nectar or pollen, or to return home. The scope of their capabilities is truly amazing; multiplied by the population of the hive, which is sixty thousand, and the world is presented with a formidable force, whose bottom line is honey!

Collecting
nectar

In a previous chapter, I said that the frames were like the rooms in a house, once the foundation was drawn into comb; and now we must look at this comb more closely. At the same time as we consider how the space is used, it is imperative to be aware of the timing involved. The crucial time is May, the month when most fruiting plants are blooming. The quantity of nectar available is tremendous and the bees nearly go berserk in their frenzy to get every drop. They visit every flower they can, rush back to the hive, deposit the load, and rush out again for another sip of sweet nectar.

Well, it isn't exactly that way. Oh yes, the honeybee does go to each flower as rapidly as she can to gather nectar; but when she rolls that long tongue down to the flower's nectary to suck up the sugary liquid, a laborious and wonderful process just begins!

The bee's
two stom-
achs: diges-
tive and
nectar
storing

As quickly as our nectar-gathering bee has filled her honey stomach, she sets her sails for home. The honey stomach is not a digestive sac but rather a collecting vessel. It is the first large opening beyond the esophagus, which is a throatlike tube. Her regular digestive stomach, or "ventriculus," follows the honey stomach, connected by a valvelike restrictor called the "proventriculus." It is an interesting fact that this restrictor allows the bee to hold her load of nectar while delivering some pollen nourishment to her regular stomach. Try not to shy away from this small bit of information about the bee's anatomy; the information here will serve you well.

During the bee's journey back to the hive, the wonderful process of turning watery nectar into beautiful golden honey begins. The enzyme invertase is added to the nectar, and then the nectar begins its long transformation into honey. Loaded with nectar, the foraging bee enters the flight pattern, waits her turn, and then lands heavily at the hive entrance. The guard bees identify her as one of their own and she is admitted to the interior of the hive. She is immediately recognized by an in-house worker bee, and there is a momentary exchange of antennae touching before the forager bee makes her offering. She regurgitates the precious cargo and then transfers it to the house bee. In a few moments the nectar gatherer will be hard at work again, but not before some quick nourishment and a speedy cleaning of her mouth, tongue, eyes, and antennae.

*Transforming nectar into honey by adding invertase*

Meanwhile the house bee heads for the first empty cell she can find where few bees are present (uncongested). She begins a process which is not unlike that of a child toying with the last bit of liquid remaining in an ice-cream soda; sucking up that last drop through the straw and then blowing it back down into the glass. Our house bee lets the drop of nectar roll out to the end of her proboscis, which is her "straw." Then she draws it back again, deep within

*Storing nectar in an empty cell*

her body. Again the nectar comes forth and again it is retracted. The house bee may continue this activity for about fifteen minutes. This is part of the ripening process, though the nectar is not yet made into honey.

After this interval of ripening activity the house bee will place the drop of unripe honey into a cell, being certain to spread it out over the bottom and side walls. The cell has a wet appearance when she finishes the first application.

*From nectar to honey takes 2 to 4 days of ripening*

This procedure is repeated thousands of times each day, and the cells holding those wet drops of nectar also number in thousands. If adequate comb space is available, the bees hang their unripe honey in the cells over a large area to continue the process of ripening. A shorter period of time is needed to ripen the honey *if* they can use *more* cells with *less* unripe honey per cell, rather than fewer cells with full loads of unripened honey.

Usually two to four days are required to evaporate the balance of the unripe honey's moisture and advance the sugar concentration to make it into honey. Once it has become honey, it is transferred to other cells or compacted where it is.

*Honey= carbo- hydrate; pollen= protein*

Honey supplies the carbohydrate portion of the bees' diet. Pollen supplies the protein. The average colony will gather seventy-five to one hundred pounds of pollen a year. The importance of pollen cannot be overestimated, since without pollen not a single bee would be born. Pollen furnishes the essential nutrients for nurse bees to produce royal jelly. It therefore becomes essential for feeding the queen as well as the brood.

Every worker bee is equipped with a special bas-

ket-like group of hairs on each of her hind legs. *Collecting*
Thousands of grains of pollen are stuffed into these *pollen*
baskets, if the worker is one of those foraging bees
detailed for that task. The function is a specific one,
not just an accidental collection of pollen while the
bee is merrily visiting the flowers to taste nectar.

The pollen collector first rolls around the flower's *Building a*
pollen-filled stamens. Her furry coat becomes dusty *pollen pellet*
with pollen as the grains lodge between and on her
body hairs. Then she hovers in the air over the
flower, while she fastidiously combs her hair with
feet made sticky by her honeyed tongue. Her forward pair of legs start the combing process on the
top of her head, gradually passing the ever-enlarging
bundle of pollen back on her body so the middle legs
can reach it. The sticky ball of pollen grows as
thousands of grains clump to form the pollen pellet,
which is stuffed into the waiting basket. If the load is
too light, she revisits that flower and others till her
"cup runneth over." When her task is accomplished
and she is ready to return with her cargo, a single
hair under the pellet pins it in place like an old-
fashioned hat pin.

The pollen-collecting bee is greeted by a house bee
in the same manner as the nectar gatherer. The pellet
is dislodged and transferred to the house bee, known
as a "pollen packer."

In a band of cells an inch or so wide and just above *Storing the*
but surrounding the brood area, the brightly colored *pollen*
pellets are pressed deeply within the cells. The pack-
ers put their heads down and like hydraulic rams
compress the pollen, using every micro-space. They
also store pollen in other areas, especially in the fall,
when they sometimes load honey in the same cell

over the pollen, and finally cap it with an air-tight beeswax covering. This pollen, stored at the summer's end, is for use the following spring. Brood rearing starts late in January and pollen, though sorely needed, is difficult to obtain at the time.

# 4

# Where to Locate Your Colony

It would seem that the location of your hive would be given some prominence in beekeeping literature. I discovered there was little or no attention given to it. A poor location can mean the difference in a colony's success or failure in gathering nectar. They'll manage somehow, but that is not what you want.

An important consideration is the direction the hive entrance will face. Experience tells you that the coldest winds and weather blow in from the north or northwest; but your experience is not likely to tell you that the sun is the bees' alarm clock. As it rises, the sun's warming rays will fall first upon everything that faces it. Of course, the sun rises in the east. We can easily combine two directions. Face the hive entrance toward the south for warmth and to the east for the sun's help in getting your bees out of bed. This means your early-to-bed bees will be early up and away gathering nectar every working day. Conclusion: face your hive toward the southeast and know the reasons why.

Mentioning the summer sun makes me think of

*"Facing" your colony's entrance*

*Sun and shade*

the plus-ninety-degree days that are made even more fiery by that orb in the sky. It gives rise to the need to shade the hive during the warmest portion of the day. If dappled sunlight, the kind that filters through the leaves, can be provided, it is of course the best solution. It is really ideal, since those same leaves won't be there in the winter and the hive will benefit from whatever warmth the sun provides. If no shade is available during midday, furnish it, possibly in the form of old venetian blind slats placed across a couple of bricks atop the hive cover. If the colony overheats, the bees are forced to carry in huge amounts of water, which is used to cool the hive through evaporation. The bees establish a relay system to carry air into the hive beginning with bees at the entrance fanning their wings. This air is moved through the hive, across cells containing water, and then out of the hive by other bees at the entrance fanning the air outward. Such a waste of energy! Each droplet of water might have been a bit of nectar. Some cooling is always required during a part of the summer, but we must try to minimize this chore for the work force.

*Level and hilly terrain*

The terrain for the hive can range from level to hilly, but if you are placing it on a hillside, neither the bottom nor the top should be chosen. At the top, the bees are exposed to the wintry blasts from the northwest. The hill's base or other lowlands collect damp, frosty air. The choice location is about one-third of the distance from the hilltop. The direction that the hill's slope faces must be given close attention. Again we look toward the south and east. Any hillside that faces toward the northwest is exposed to the fierce winds of winter. Placing bees in such a vulnerable spot can only end in disaster.

Optimum conditions are rarely available to us. *Too much* Sometimes we must make do with what is given. *cold wind* Perhaps this is true; let us see! I have a bee yard of thirteen colonies that breaks almost all the rules. They are situated on a hillside that slopes directly to the northwest. (Please remember, that is the *wrong* direction.) The hives are set in a semicircle. The hive entrances face south and east, except for three. They face the west; they are the opposite leg of a horse-shoe. Now, with that setup, you would imagine I would know what to do. Well, I knew and should have moved them, but I didn't do it. Instead, I bought a huge roll of burlap. Then on the north and west sides of my hives, I post-holed myself into exhaustion. Those post-hole digging tools are not kind to the human animal. Next came the posts in the form of new-felled saplings that were to be cleared anyway for the new barn. Out came the chain saw and I had the posts in place in a line about twenty feet from the hives.

I grappled with the burlap, unrolling it along the posts and some trees which offered their help by being there. My staple gun was in a hip holster, and so, as each post and tree was reached, the burlap was secured. Temporarily! The first windstorm showed me that staples were not the way to hold burlap in place. I considered the job completed when I sand-wiched the burlap between the posts and inch-wide strips of wood nailed for the entire width of the material. Halfway between this outer fence of burlap and the hives, I ran a second barrier in exactly the same way.

Well, winter came, stayed awhile and left. It was a

rough one. The bees had plenty of stores going into the height of winter, so that did not concern me. They had adequate ventilation to keep them dry, and they were strong in numbers of young bees. But oh! did the wind blow that winter from the northwest! On the first mild day in late February, I peeked. I don't think I know of anything that tears the heart out of a beekeeper more than looking into a dead hive. I had to look into three such hives. The three facing west were victims of that winter's chill wind. There was food aplenty, but it was apparent the bees were unable to reach it owing to the continuing cold. They starved to death. The other colonies made it; though on the wrong slope, they faced away from the wind and the burlap furnished the windbreak.

*Water* Several other factors combine in importance to influence the final selection of a colony's location. Water availability is a prime consideration. Your neighbor's swimming pool must not be the closest source of water. If it is, you must be diligent in supplying water to the hives. Usually a chicken watering device is the simplest means of doing so. They are inexpensive (under two dollars) and easy to use.

*Moving a hive* You must also locate the hive with an eye to the harvest. I know you probably feel we are looking too far ahead, but decision time is now. Once the colony is established, it is extremely difficult to move it to another location within the bee yard. Moving it two or three miles, or two or three hundred miles, offers no problem; just try to move it thirty feet and watch what happens. All the bees will leave the hive normally, to forage; but when they return they will go back to the former location and mill about, wondering, "Where did the house go?" Unfortunately, unless assisted these bees will eventually perish from exposure.

Don't place the hive somewhere that a wheelbarrow or some similar vehicle cannot easily reach. Even a shallow super of honey can weigh sixty pounds. Wrapping your arms around sixty pounds of honey and trudging up a hill or through the woods is no one's idea of fun. So make it easy to roll a wheelbarrow or—even better—a four-wheeled vehicle (your son's wagon) to your honey super.

You really do not want your colony in the woods anyway. It has been proved that such colonies are usually cranky, irritable, and feisty. It is probably the complete lack of sun that does it.

If you put four fifteen-inch legs on your hive stand, you will succeed in avoiding several annoying problems. At once your back will thank you, since you will never have to bend from the hips and creak. The top bars of the frames in the first deep body will just about meet your fingertips. *Put legs on the hive stand*

Skunks have a yen for honey. They will approach a hive at night, scratch at the entrance, and when the bees investigate, friend skunk will grab a handful. After he has chewed on the bees to get the honey, another helping usually follows. Repeated visits of this type can make a colony irritable and unmanageable. Frequently a beekeeper wonders why his normally docile bees have become more difficult. At any rate, your standing hive stand corrects this situation. When the skunk reaches up to the entrance, his underbelly is exposed to the bees. This is the skunk's vulnerable area, and the bees know how to handle it. Evidence of a skunk's visit are telltale *Skunks*

A perfect location

clumps of expectorated bees on the ground.

*Over-*
*colonizing*  Usually an enterprising beekeeper may be concerned with another aspect of location which I have not mentioned, the potential yield of the area. I haven't because a hobbyist beekeeper with one or two colonies, or even as many as a half dozen, need not concern himself with the nectar volume available from his given area. It is nearly always in excess of the bees available to use it. Flowers go begging for someone to sip their nectar. Naturally, if one prospers and adds additional colonies endlessly, an area

can be too heavily saturated with bees. A friend of mine decided to go semi-professional and installed three yards of forty colonies each. The first and last were about twenty minutes' driving time from each other. It wasn't until he checked the locations on a map that he saw the yards were only about a mile away from each other. That meant the bees could meet at a point a half mile from their hive. All three yards were within two miles.

He also set up too many colonies at one yard. Again we must consider the area. If the location was a two-thousand-acre farm with a variety of crops, that might be all right. If we are talking of suburban and urban locations, the number of colonies in one yard should be limited to fifteen to twenty. One need not compete with himself.

A last and important concern, but truly not the least for the hobby beekeeper, is other people. Never aim your hive at passerby pedestrians, stables, or kennels. Keep clear of horse traffic and cowpaths. If a colony is located at least one hundred feet away from any vulnerable human or animal, you need not be concerned. If you are forced into a "tight" location, provide a fence, or a row of trees, or even a hedge, some ten to thirty feet away. It will force the bees to fly up and over the heads of people and animals. This much consideration on your part will earn for you the respect and friendship of non-beekeepers. The opposite can only gain a "nuisance" reputation.

*Concern for others*

# 5

## The Day Your Bees Arrive

*Ordering your bees* Once you have assembled and painted your first hive and selected a good location, you are ready for your first tenants. The time for starting with a colony of bees is spring—early spring if you live in the South, May or June elsewhere. And you should order your bees as early as possible. There are shortages, and late orders sometimes bring disappointment. February is not too early to place your order. Bees can be ordered through your local bee-equipment supplier, or by mail from ads in any of the beekeeping publications. They are raised commercially for sale on bee farms in Georgia, Alabama, Louisiana, and other southern states, and are shipped by mail in wooden boxes with screening on two sides. Several-size packages of bees are available, from two to five pounds. Three-pound packages appear to be just the right size, while two-pound packages are a little light.

*Different types of bees* The type of bees you order is up to you: Italian, Midnite, Starline, Caucasian, Carniolan, or any other. Each kind has advantages and some disadvantages. A top honey producer, resistant to disease and very gentle, is the kind most highly desired. A great amount of hard work has gone into producing hybrid bees that can meet some of these standards. Geneticists have been trying for a number of years, and it looks like they are reaching their goals. But a beginner is best advised to start with an established strain.

One morning, as early as six o'clock, your phone will ring. The voice at the other end will say, half pleadingly, half demandingly, "Please come and get your bees." It will be the post office and you will have been waiting for this moment. Unless previously educated by some other beekeeper, the caller will be anxious to dispose of this "hot" cargo. He needs your help. While on the phone ask him to put the bees in a cool, shady place—if necessary outside the post office. Trying to be helpful by keeping the bees warm can result in disaster. The call to pick up bees does not always come at the early hour, but it seldom comes at the right time.

At the bee farm when the bees are "shook" from their original hive into the shipping cage and closed up, a few bees will end up on the outside of the screen looking in on their caged friends. They are called hobos. They will ride along and never part from their sisters unless brushed off. These few bees are generally the reason for the sound of urgency from postal employees. They may think that the cage has a hole and the bees are escaping. If this is the case, enlighten them to the facts and assure them you will be there posthaste. Prior to this moment, it would be prudent if you were to inform your wife or husband or some other person at home what to do in case you might not be available. Just be sure that when the post office calls, your response will be as immediate as possible.

At the post office, inspect the bees, and unless there is a depth of more than approximately one inch of dead bees across the bottom, take them home without delay. They have had a tough trip and need your attendance.

If it appears that your loss of bees approaches half of the total, then write up a statement spelling it out. Ask the postal attendant to sign it for you. Bee farms usually insure the bees and will replace them under these conditions. They must have the signed statement when notified. You will pay the new shipping charges if the package is replaced, but not for the replaced bees. I should mention at this point that later, when you open the package, should your queen be dead notify the bee farm and she will be replaced without any further charge.

If all went well with your bees, head for home. Place them on the seat inside the car, not in the trunk! Make certain they do not remain in the sun for any long periods of time, even with air conditioning on.

As soon as you reach home, have a nice cool glass of water. One for you and one for the bees. Yes, your girls are hot, tired, and thirsty. Move the package to a cool, rather dark place, such as your basement or garage. Next, spray the screens of the cage with tepid to cool water. Your sprayer should be new; but if you won't splurge, be sure that only water has passed through the one used. Don't poison your work force by error, using the garden sprayer or mister that applied fertilizer or insect spray. If you don't have a sprayer, use a clean paint brush; dip it in water and apply to the screen. (Care must be taken not to injure the bees' feet clinging to the screen.) The water will cool the bees and quench their thirst. Spray them on all sides. When they look wet, it will be enough. This is the moment to begin learning a new language. As you gradually enter the bees' world, you will notice that they have their way of expressing pleasure and displeasure. While imprisoned in the cage, they will "ask." After their drink, you'll notice how quiet it becomes. Just before you sprayed them, there was an uproar. In a couple of hours there will be a clamor for attention again.

If you are a purist, you might skip over the next important bit of information. I can't risk having you do that.

*Feeding the bees sugar*

It has to do with feeding the bees a syrup made from white granulated sugar. Wait, don't run away till I tell you that the sugar never, never goes into or becomes honey. The syrup never adulterates the honey when used for spring and fall feeding. Sugar syrup and/or the nectar from the flowers provides the raw material for the bees to make wax. In the beginning, you feed it strictly for the production of wax for comb building. A small amount is consumed by the bees as the carbohydrate part of their diet. In the fall of the year, after surplus honey has been removed, it is again fed, but this time to supplement stores of winter food. Again only consumed by the bees.

*Measuring water and sugar*

After the bees have been given their drink of water, use the ensuing period of quiet to prepare for their next spraying. This time you will note that your gang won't be satisfied with water alone; make up a syrup by mixing five pounds of granulated sugar with five pounds of water. Note that we figure our mix by weight. A gallon of water weighs about eight pounds. There are four quarts to a gallon, so each quart weighs two pounds. Carrying this a little further, so you'll be prepared for all seasons: there are two pints to each quart of liquid, each pint of water weighs one pound. My mother used to say, and I suppose millions of other cooks, "A pint's a pound the world around."

*Mixing the syrup*

Now as we start measuring and mixing, pour five pounds of hot water (two quarts plus one pint) from the faucet into a two-and-a-half-gallon plastic pail. If you must heat your water on the stove, be sure and use a metal container, not plastic, and don't let the water boil or put the sugar in the water while it is on the stove being heated. (Honey bees get terribly sick from burned or caramelized sugar.) Make sure the pail and other equipment are not contaminated by being used for other purposes. Now slowly pour the five pounds of sugar into the hot water, not vice versa, if you wish to save time, stirring as you pour in order to dissolve it completely. The liquid syrup should become crystal clear with stirring.

*Use one-gallon glass jars*

Pour this mixed gallon (that is just about what you get from a one-to-one solution) into a one-gallon glass jar, preferably with a wide mouth. These jars can often be found at restaurants, which normally dispose of them. They usually have contained mayonnaise, fruit segments, pickles, etc., so a good washing is necessary. If detergent is used, be sure to rinse thoroughly. Most of the jar caps these days are without a waxed insert for making a tight seal. Most jar caps come with a formed plastic gasket which is an integral part of the cap and makes a tight seal. (If you use jar caps with a waxed insert for a seal it is necessary to cut a two-inch square out of the insert for the syrup to pass through.) That seal will allow you to invert your jar of syrup, after you have made five or six tiny holes in the metal cover, from the outside to the inside. Use a very small nail and allow only the point to perforate the cover. When you turn this jar upside down, there should be no leakage from the holes after approximately five seconds, because of the vacuum formed inside the jar. Only a tiny bee tongue inserted into one of the tiny holes will break the vacuum and allow syrup to come out.

Before I tell you where and how to place this jar for the bees to use, I advise you to add a small amount of preventive medication to the syrup to guard against certain bee diseases. Incidentally, these diseases have no effect on the honey or on human beings. The bees alone have the problem.

*Adding bee medication*

Adult bees can and do contract a few diseases. Most are not serious, and the bees manage them just as we handle a common cold. Nosema is a common disease sometimes brought on by "package bee stress," which affects them just as a cold affects us. During and after a cold or flu, if we don't contract something else because of lowered resistance, we usually don't feel too energetic. We rarely feel much like working or being productive under those conditions. So it goes with our bees when nosema apis strikes. It can hit a colony viciously, killing some of the bees and seriously debilitating a hive in short order. While it rarely will kill an entire colony, it will hinder a colony's development. Generally one doesn't find out until it is too late. The weeks fly by from mid-April or so when you hived your newly arrived family. Suddenly it is June first and your group just doesn't seem to get up to speed. The

*Nosema apis*

flowers bloom, the nectar spurts and your bees are not getting their share of the yield. They might have that "cold" (nosema) and never multiply their numbers sufficiently to provide a foraging force of bees. Such a colony might not develop sufficiently to make it through the winter for lack of accumulated honey stores. The honey is necessary, as fuel to energize their muscles to provide heat within the cluster during the long winter wait.

*Use fumigillin*

The way to avoid this problem is to use fumigillin, known commercially as Fumidil B. Most bee-supply dealers stock it and generally have it available for spring and fall feedings. One level teaspoon dissolved in your gallon of syrup and repeated in the second gallon you give them after hiving the bees, will help prevent nosema from developing and cure it if the colony is infected.

*American foul brood*

Another dangerous disease that makes nosema look like a kitten in comparison, is American foul brood (AFB), carried by a spore-producing bacillus. This disease destroys the new bees in their larval and pupal states, which are pre-adult. This means the colony will die, since no new bees will ever be born. It is a virulent disease and very contagious. Even after all the bees are gone and the hive is cleaned out, the equipment used remains as a source of reinfection for any new bees put into it. This holds true if the bees are replaced immediately or five years later. The reason is simple. A spore, which is a thick-walled bacterium in a resting stage, is left behind after the bees have died. It sits dormant, safely ensconced in its fatty covering, waiting for a new host to appear. When it does, the process is repeated and the new bees fade away again . . . and so on ad infinitum.

The disease is communicated from colony to colony by exploratory scout bees from strong, healthy colonies seeking sources of nectar and honey. The scout bee generally manages to get into only a weakened or small colony. If the weakened colony is infected, the scout brings the infection home via the stolen honey, in which spores may be present. This is the reason why the activity known as "robbing" must never be tolerated or caused by you. The ugly truth is that if your strong hive (a year from now or less) should have even one foraging bee visit a diseased colony (or the remnants of one) there is a good chance your colony will be infected. Used equipment hiding in an old barn can be a good source of disease.

You can avoid American foul brood in the beginning by adding to the same gallon of syrup, along with the Fumidil, one-quarter teaspoon of sodium sulfathiazole. A bee colony, fortified at the outset by Fumidil and sodium sulfathiazole, will grow strong in numbers and better cope with the rigors of nature.

*Use sodium sulfa- thiazole*

The best way to add both medicines is to mix up one teaspoon of Fumidil and one-quarter teaspoon of sulfa in a little *cool* water. The chemicals are easily dissolved in a closed jar of any kind containing a couple of ounces of water. Shake the jar violently and the powders will dissolve. Then add it to your gallon of syrup, mixing it in thoroughly. It is wise to have this accomplished before hiving your package of bees. Make up the jar of syrup and carry it with you when you take the bees to the house which becomes their home.

*Mixing in the medication*

We have now reached the traumatic moment that *every* beginning beekeeper experiences. At this point

*Preparing to hive your bees*

it helps to remember that your ten thousand bees (three pounds) have only one objective, to get inside that lovely house you built for them. Once inside, they will vigorously attack the foundation and make thirty-five hundred rooms (cells) on each side of each sheet of wax. Then their magnificent multiplying machine, the queen, will begin to lay one thousand or more eggs per day. The queen is all important, but those laboring workers of the hive must first engage themselves in the comb-building process before she can lay any eggs.

*Quiet them with syrup*

Just before H hour by thirty minutes or so, go to your newly acquired charges and give them a good feeding of sugar syrup by spraying them through the screens on both sides of the package cage. (Or use a clean brush dipped in syrup.) They will enjoy cleaning each other off. All will become fat and happy for the next few hours. They also will find it difficult to fly while wet and sticky, and that is the way you want it.

*Assemble your equipment*

Check out your hive, positioned with the entrance facing south or southeast, *slightly* tilted forward but absolutely level from side to side. Your equipment should include a section of newspaper, hat with head veil, gloves, hive tool, spray bottle with syrup, gallon jar with syrup, pliers, and a hammer. Also have two frame nails, each bent in the middle to a right angle (ninety degrees—like a right-hand turn when driving). When all checks out, you will fetch the future residents of the hive and an upper hive body, but without frames. Forget about the smoker; it is of no use at this time.

Within the next five minutes, all the preparations you have made will come together to provide the newest experience in your life. I have hived the homeless hundreds of times and never have failed to feel elation, a tremendous surge of well-being. This first encounter for you is filled with trepidation, even a little fear. Each time it is a little like that for me, for aren't we spilling out ten thousand creatures, each with a valiant lance? Fear not; instead enjoy and thrill to the creation you are bringing into existence.

All right, get started! First a deep breath! It will be finished in a few minutes and you will probably say, "How simple and what an interesting experience." It is a first step into a new world that will promise a new experience you won't dare tell anyone about.

*Take a deep breath*

Use your hive tool (flat end) to raise the square board nailed to the top of the cage. Raise it enough so that when you push the board back down the nail heads will be convenient to pull out with hammer or pliers. Under that board is a feeding can of syrup for the travelers. Next to the can, in a separate slot, is a small queen cage housing the queen and a few attendant bees who feed her. Pull the nails, but don't remove the board.

*Raise the package's board*

Remove the inner and outer covers of your new hive, leaning them against the back of the hive. Remove the upper deep hive body (food chamber), if you had it on top of the lower deep hive body (brood chamber). Empty it (upper hive body) of all the frames, which you will store for a few weeks, while the bees expand the brood chamber. Remove five frames from the brood chamber. Lean these five frames against the cover; you will use four of the five shortly.

*Prepare the empty hive*

At this point, you have an open hive with five frames all pushed together on one side. Place your

*Open the package*

package of bees atop the frames and spray them with syrup again, on both sides. Use judgment; wet them down, but don't make the bees into a soggy mass. Place your right hand over the square board you previously loosened, palm down, with your thumb on the screen on one side and your four fingers on the other side. Raise the package an inch or so and then jar it down, causing the bees to drop to the bottom. Remove the previously loosened board, placing it on the ground within easy reach. Your objective is to remove the syrup can and the queen cage before the bees come up from the bottom. Don't rush yourself. Removing can and queen cage only takes a

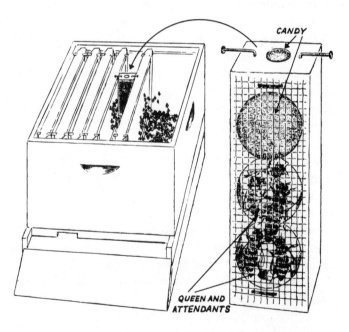

The correct position for the queen when hiving the bees

few seconds, and the bees remain below for nearly half a minute.

Removal of can and cage comes in that order. Place your index finger on the disk over the top of the queen cage and your thumb as before, along with the remaining three fingers on the opposite side. Support the entire package in your left hand with your right hand on top. Make a quick upward movement and the food can will rise into your right hand. Once you've grasped the can, remove it. Put it down anywhere convenient. It has a hole in the end that pointed downward in the package cage. There may be a few bees on it; just ignore them. The queen cage is now easily removed. Place the square board back on the opening of the cage.

*Remove the package syrup pan and queen cage*

Now take another breath. Don't forget to breathe. All the bees are contained in the large package cage and the queen is now ready for you to check. Is she alive? Yes! O.K., continue. If she is dead I'll tell you what to do later. Don't be concerned over dead attendants with the queen; it is a normal situation for some or all to be dead.

You now have work to do on the end of the queen cage that has the white candy. Either a cork or a metal disk will cover the hole at that end. When you remove either, and you must do so, go slowly in case all the candy has been consumed. You could lose the queen unless she has had a wing clipped. Queens can fly quite competently. Remove the cork or disk and you will see the candy. If the candy is not there and you can see to the inside of the cage through the hole, close the hole temporarily and go to your kitchen and make up a candy plug. Ordinary confectioners' sugar and water mixed into a dough will

*Insert the queen cage in the hive*

suffice. Plug the hole to a depth of at least a half inch. Use a small-diameter nail (one of the bent frame nails will do) and push a hole clear through the candy. Careful, don't spear the queen! Using your pliers drive the nails into the queen cage as the sketch shows. These nails become the hangers for the queen's cage. Place the cage between the frame nearest to the center and the one next to it. The screen on the queen's cage should face toward the center. Position the whole cage a couple of inches back of center. The screen does not face the frames, and the candy end faces upward. This eliminates the chance of dead attendant bees blocking the queen's exit, once the hive bees have consumed the candy. The bees are originally attracted to the candy because

"Inside feeding" spring and fall

the queen's pheromones pass through the hole that you put there with the nail. The hive bees feed the queen through the screen during the couple of days it takes them to clear away the candy.

At this point the queen is in position, the food can is set aside, and frames and covers are leaning against the hive ready for use. (If you discovered a dead queen, use the cage *with* the dead queen anyway. Notify the breeder for replacement.)

*Pour the bees into the hive*

Raise the package cage and give it one smart rap on the ground. Now reverse your hands. Place your right hand underneath the cage and remove the square covering board with your left hand. You can throw it away now because you're finished with it. Spill the bees, as though you were pouring liquid, through the round hole and over the queen cage. After about half the bees are out, spill the rest into the area where there are no frames. When only a few bees remain in the cage, jar it down on the bottom and the side, to accumulate the bees in one corner. Next, roll the bees out of the hole and into the hive.

Keep your attention on what you are doing. Any flying bees are not looking for you. They are intent upon ridding themselves of accumulated feces and then getting to work on that nice-smelling wax in preparation for the queen's work. Lay the cage on its side in front of the hive so any bees still in it will get out on their own.

*Replace four frames and put on inner cover*

Don't skip the next step, though you will be tempted. It is the next to the last move in the hiving procedure. See if the bees are still piled up higher than the bottom bars of the adjacent frames. If so, use your gloved hand to gently smooth out the bees. Spread your fingers wide and make believe your hand is bare. Be very gentle and don't push your fingers to the bottom of the bees; just comb over the surface. Now you can place four of the five frames back into the hive. The fifth you'll use later when the queen cage has been vacated and removed. Replace the inner cover by gently sliding it on very slowly. You will not kill any bees this way. The ledge side of the cover faces up.

*Add inverted jar of syrup*

Pick up a couple of twigs about one-eighth inch in diameter and as long as the oval hole in the flat section of the cover. Place them on each side of that hole. Invert your gallon jar of syrup over the hole but resting on top of the twigs. The space under the jar cap prevents squeezing bees. It also provides the bees access to the syrup holes.

*Add upper hive body and insulation*

Position the upper deep body, without frames, on the inner cover. Close the ventilation hole in the rim of the cover with the half-moon block that comes with the hive, or with grass. Surround the jar with crumpled newspaper, filling the entire hive body. Top it off with several thicknesses of a flat section of newspaper. Replace the outer cover and weight it down with a good-sized rock. The newspaper provides insulation for the jar of syrup. The jar's cover is warmed by the heat rising through the oval hole. The cover, in turn, warms the syrup just before the bees get it.

*Insert an entrance reducer and entrance feeder*

If you placed the entrance reducer in the hive's main entrance before you started, you won't have to do it now. Use the largest opening in the reducer. Insert an entrance feeder into the opening. This will leave a very small entrance hole, about enough for one bee to pass through at a time. The restricted opening serves two purposes. The chance for other bees to rob your new and defenseless colony is eliminated; and it helps to keep your bees home during the first couple of days following the hiving. Important! Do not open the hive for a week. LADIES AT WORK. DO NOT DISTURB.

During the next couple of days, if you have the opportunity, find a rock to sit on close by the hive. Give yourself a few sweet moments while you get acquainted with your bees. Sit to the side of the entrance within reading distance. See the pollen on their hind legs as they return from afar. Watch them shoot out of that entrance, as though jet propelled, on their way to forage for nectar. Enjoy the beginning and it will never end.

*Watch and enjoy*

# 6

## The First Few Weeks— a Critical Period

The few weeks following the installation of bees in a new hive is without question a most important span of time. During this period your colony will realize an amazing number of accomplishments. Their main task is to rear young bees until their number is fifty to sixty thousand strong. Remember, they are now only ten to twelve thousand bees . . . and a queen.

*Drawing out combs for egg cells*

You gave them a house; but it is they who must furnish it. Each sheet of wax foundation will be expanded into a comb, which will about equal the width of the frame's top bar. In order to "draw" out the foundation into comb, the bees make a huge amount of wax. Just as the cow has milk glands, so does the bee have wax glands. It is only the female worker bee who can extrude these wax scales from plates on her underbelly. These particles of wax, which look like minute fish scales, are shaped and added to comb. The building process continues until each frame has seven thousand hexagonal rooms.

In the beginning the rooms will house tiny eggs

laid by the queen in ever-increasing numbers. As soon as a few hundred cells of comb are built, the queen will fill each of them with a single egg. Though the eggs are quite small, they can be seen clearly, each resembling a tiny grain of rice perched on its end. As more and more cells are built, the queen will race on, depositing more and more eggs per day.

*The furious pace of egg laying*

In the first days, a hundred to three hundred eggs are deposited each day depending on the space; but each succeeding day the queen seems to develop a furious pace. She will fill every newly constructed cell even if it means laying as many as two thousand eggs every day. The bees do their best to accommodate the queen and frantically attempt to provide the room for the egg laying. They seem to be involved in a race, the bees and the queen. Make room, lay eggs. Make more room, lay more eggs.

*Help them with sugar syrup*

You must be involved in the race, for where can the bees get the fuel to make the wax, to make the rooms for more eggs from the queen? Why, of course, from you. It is usually mid-April when most of us start a new colony of bees. Very little is blooming so early in the spring; certainly not enough nectar is flowing to feed the bees' wax-production machinery. Without a supply of nectar this mad scramble to increase their numbers slows down and finally comes to a halt. So feed your bees and don't stop, if you want a strong colony prepared to reap their harvest from the blooms of summer.

*The crucial timetable of bee gestation*

At the same time that all of this is happening, other events are taking place. Remember, we are discussing "the critical period"; let me tell you why! Your bees that came in the package are, hopefully, about a week or so old; in other words, young bees. Some might be three weeks old. The life span of a bee during the active part of the year (spring through fall) is only six weeks long. If you hived your bees on April 15 and it is now April 29, the original young bees are now about three weeks old; roughly half their life is gone. Oh, yes, the queen started laying eggs shortly (about a week) after you hived the package, so there are some bees on the way. It takes three weeks (twenty-one days) from the laying of an egg to the birth of a worker bee. Let us make a small calculation to see what is really happening.

1. You hived the package on the 15th.
2. The queen started laying about the 22nd, at the rate of 200 per day for the first three or four days. (We'll get back to inspecting the hive for this occurrence in just a little bit.)
3. About the week of the 29th the queen got up to speed (provided the bees made space for her) and she deposited 1,000 eggs a day.
4. By the *end* of this same week of the 29th, your package bees are at least three weeks old.
5. Around the 13th of May (3 weeks after the queen laid eggs on the 22nd of April) about 200 bees will be born. Each succeeding day 200 or more bees will hatch. On the 13th of May your package bees are five weeks old and have only one week left to live.
6. The crucial week arrives: the 20th of May. During this week, nearly all the package bees will die. They are replaced by the bees born from the eggs laid during the week of the 29th of April. The 29th of April was the beginning

of the third week you had your bees, when the queen began laying 1,000 eggs a day. So, beginning on the 20th of May, a thousand bees will be hatched each day.

You see, your colony will dwindle down to a precarious low just before it begins to grow—and grow it does; almost like an explosion, for at that vulnerable bottom, there are usually about 30,000 to 40,000 bees in various stages of development. Some are only eggs, remaining so for two days; others have hatched from the eggs into tiny larvae. This worm-like stage will change after six more days into another phase of development. It is the pupal period, which precedes the hatching of the adult bee. During this last interval, the pupa is enclosed in the cell by a flat wax capping. It is called "capped brood." And at the end of six weeks, it is comforting to see five or six frames filled with developing bees; most frames should have plenty of capped brood—at least a third of each side of a frame—easily identified by the capping.

*From egg to larva to pupa to bee*

It becomes obvious that during the early development of a colony we must give it close attention. Later we can let them shift more for themselves. A steady supply of nectar or syrup will develop the comb quickly. If we have an efficient queen, she will lay eggs diligently.

Is every queen a good and efficient queen? The answer to that question determines how strong your colony becomes, within a given length of time. Before we learn how to qualify a queen, we had better go back to the hive and make sure she has been released to go to work.

# 7

## The First Week

After hiving the bees, positioning the queen cage, and setting up their food supply, it is prudent to leave the bees alone for about a week. At least five days must elapse, unless there is no activity and it becomes imperative to inspect the bees sooner (see page 80). The touchy part of all this relates to the acceptance of the queen by her bees. Somehow they feel that any interruption by you must be the queen's fault. So they mass around the queen, forming a ball, killing her through suffocation or starvation or perhaps both.

However, you may replace syrup-feeding jars as required. In fact, this must be closely watched. As soon as the syrup is consumed, replace it. Keep up this feeding until all the frames in the brood chamber have fully drawn comb available for the queen. Medication in the syrup should be discontinued after the second gallon has been fed; but continue feeding.

*First-week activity* During the first week, if weather permits, the bees will bring pollen into their community. Depending upon location, one or more early blooms will provide pollen for the new colony, particularly the willows.

Watch the returning bees carefully. You can see the little colored pellets of pollen on their hind legs. As I have said, pollen supplies the protein portion of the brood diet. It is absolutely essential and relates directly to a quick buildup of the population.

I have made it a ritual to feed pollen patties to all colonies each spring, whether new arrivals or old, established colonies. Pollen substitute, composed of brewer's yeast, soybean flour, and dried skimmed milk powder, is available as a prepared mixture. This blended powder is combined with a quantity of pure pollen powder. The powder is then mixed with sugar syrup to make a doughlike mixture (see page 178). It is rolled out between two pieces of wax paper to a thickness of about one-quarter inch. A strip of patty six inches long by three inches wide may be placed directly upon the top bars. A few holes in the wax paper on the underside will enable the bees to gain access to it more rapidly. The wax paper on top will keep it moist and the bees will eat it all, even the wax paper. A constant protein supply helps to develop strong colonies; and a strong colony is one that has a numerically superior force of foraging bees at the time when the flowers are blooming and throwing off (secreting) nectar. This is the material that the bees must have to make honey.

*Pollen patties for protein and carbohydrates*

### The First Look Inside

Having impatiently waited out the week, you can now get to the hive, open it up, and see if the queen is okay. But first make sure conditions are right. When the eggs are laid, the bees keep them warm,

*When to open the hive*

slightly more than ninety degrees Fahrenheit. Even the larvae and pupae are kept at that temperature. Now, if it is a clear, sunny day with temperature in the upper fifties or better, and *no wind*, a quick look at a comb of eggs is permissible. If any breeze is blowing, postpone the look into the hive's frames; but you do have a job to do. The queen's cage must be removed and the tenth frame added to the other nine.

*Start smoker, put on hat, veil and gloves*

Sometime between 10 A.M. and 2 P.M. begin your inspection. Start by building a fire in your smoker in the manner described earlier. Once you get the chimney going proceed to the hive with hat and head veil on, and with hive tool and smoker. Yes, wear gloves if you feel you must. Eventually, you'll work the bees without them; but use them, if you wish, in the beginning.

*Approach from the side or rear*

Opening a hive follows the same procedure each time, regardless of the reason for the inspection or visit. Stand to the side of the hive, approaching it also from the side or rear. Do not walk across or into the front of a beehive for obvious reasons. Watch the bees for a moment to see which way they are leaving and returning. Sometimes they fly directly to the right from the entrance, sometimes directly left. Most often the bees have a straight runway takeoff before turning. It is the sometimes right or left you must watch. If they leave to the left, position yourself on the right side of the hive, and, of course, vice versa. (It is remarkable how many times I have been told by new beekeepers how their bees attacked them as they walked up to the hive.) Now, from the side and with the smoker nozzle a couple of feet away, squeeze the bellows, forcing three good puffs

Smoke applied under outer cover

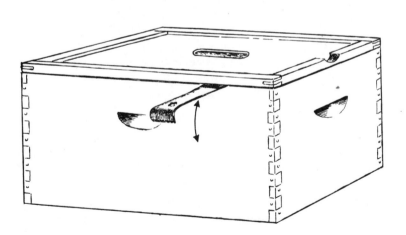

Slipping hive tool under inner cover

of smoke into the entrance. This smoke is for the benefit of the bees on guard duty. Remember, it is smoke you want, not fire, so keep the nozzle about two feet away.

*Use the smoker* Lift the side of the outer cover nearest to you a trifle and puff some smoke under the overhanging edge of the cover while holding it up. The smoke will creep between the two covers, gradually sinking through the oval hole in the flat section of the inner cover.

*Remove outer cover* While you still have hold of the outer cover, raise it higher on one side, until you can just see the upper edge of the inner cover. Now you can push it away from you so the edge of the outer cover is resting on the edge of the inner cover nearest to you. If you just pause for thirty seconds or so, the bees will have an opportunity to react and head for the nectar. Now you are ready to remove the outer cover. Raise the edge again and puff once or twice underneath it. This takes care of a few bold guards who remained on the inner cover. Set your smoker down nearby and with *two* hands lift the outer cover straight up very, very slowly. Turn it over as you place it on a flat spot on the ground. Its top is resting on the ground and the overhanging edges are facing upward.

*Use outer cover as a ground base for hive body inspection* Develop proper habits now and you will be a good beekeeper later. As your beehive grows in size, you may wish to remove a hive body for an inspection of the frames. Use the inverted outer cover as a base, placing the hive body across it. Only the four crossed edges will make contact, and few if any bees will be squashed.

Naturally, in the early stages you will tend to open

and close the hive with a few extra steps not included in the previous text or later. For example, removing the outer cover while inside feeding is in progress will reveal newspaper and gallon jar inside the extra upper hive body. You will need a carton to hold the crumpled newspaper, which must be removed along with the jar and the empty hive body, before you can remove the inner cover. A *small* amount of smoke, one or two puffs directed through the flat oval hole in the inner cover, as the jar is removed will quiet the bees.

*Remove newspaper, jar, and protective hive body*

Insert the flat end of the hive tool between the inner cover and the top edge of the hive body about three inches back from the front of the hive. You should still be on the same side of the hive. As you pry these parts from each other, place your other hand on top of the inner cover, above the hive tool, and press down. This prevents the inner cover from popping up with a snap and crackle that alerts the whole colony to your coming. They don't like sharp sounds and vibrations.

*Using the hive tool to loosen inner cover*

Once a crack is made by the tool, blow some smoke into the hive at that point. The bees can't come out, but the smoke can go in. During a ten-second wait after smoking through the crack, move around to the other side and loosen the inner cover in two places, toward the front and toward the rear. Go back to your original side and again insert the hive tool between cover and body, but this time toward the rear.

*Use more smoke*

Now you are ready to raise the cover. Lift it as you would an old-fashioned cigar-box cover, or like the lid of a car's trunk. Lift it slowly from your side as though it had hinges on the opposite side of the

*Raise inner cover and set it on end on ground leaning against hive away from you*

hive. When it is raised about halfway toward verti-cal, take it off completely. Become accustomed to clusters of bees on both sides of the inner cover while going through this operation. Place the cover down gently so one end rests on the ground, with the opposite end leaning against the side of the hive away from you.

*Use more smoke*

Bring your smoker into play, using a slow, easy motion. Point it at the top of the frames, from the rear of the hive. Three or four light puffs from about two feet away and you are finished with the smoker for at least five to ten minutes. Take note of each reaction the bees display as you smoke them. You'll establish a relation between your actions with the

The proper smoking procedure on frame top bars

smoker and the bees' reactions to it. You will learn how to use just enough smoke.

Before proceeding, determine that you will always look closely at the point on the hive or its parts where you will next place your hands or hive tool. It could be called foresight; I choose to think of it as gentleness.

One day, several years ago, I came roaring into our kitchen with a flock of bees not far behind me. In fact, I still had some on my back. After my wife had brushed them off, I sat down and began to wonder. Several stings had me thinking that I had apparently done something wrong. Bees never sting without cause. As though reading my mind, the bee mistress asked me what had started the uproar? I thought about it and realized that, while removing a frame quite slowly and gently, it "hung up" just as the bottom bars were about to be exposed. I twisted the frame trying to free it. After wrestling with it for a few moments, my patience ran out and, instead of lowering it back into the hive, I gave it a tremendous tug. My hands flew into the air with the frame and so did a shower of very angry bees. My response is usually expressed as a "hasty retreat."

*A lesson in foresight and gentleness*

As I finished my description of the event, Anita looked at me with her wisest look and dropped a real gem in my lap. "I suppose you realize your working force are all females," she began; I bent my throbbing neck in agreement. "Well," she said, simply, "females respond to gentleness; you should know that."

Ever since that time I have practiced a new form of gentleness, interwoven with compassion and understanding. These elements have allowed me to see all of nature with new understanding. Now I'm rarely stung by my bees.

*Removing the frames for inspection*

Now that you have reached the inside of the hive, you must accomplish certain tasks in as short a time period as possible—but contradictorily, without rushing. Insert your hive tool's curved end, sharp edge down, between the frame nearest you and the next one over. Locate the hive tool between the top bars and just a bit in from the end bars of the frame. A slight twist sideways exerts tremendous leverage and the frames will separate. Do the same twist at the other end of the frame. Pick up the frame by the top bar ends. Look before placing your fingers on it, and if there is a bee in the way, push her gently to the side. Lift the frame straight up and a little toward you. This will avoid rubbing bees on your frame against bees on the adjacent one. At this early stage of your colony's development, there will probably be nothing to see or do on the first frame.

*The importance of the first frame*

This next step, as all the others I have explained, is practiced each time you open a hive. When you remove that first frame, you have given yourself a slot near you, to move each succeeding frame into, before raising it for examination. Never pull a frame out until you have made room by removing this first frame. But where to put that first frame, and how, becomes more and more important as the colony grows. Consider that later on it will be covered with bees, at least on one side.

*Rest it on the ground, leaning against the hive*

Let one hand hold the frame so that the long dimension is now vertical. Hang on to it, with your hand holding the joint of the top and side bars. Rest the frame vertically against the hive with one end of it on the ground. You will find that it is touching the ground and the hive at only three points. On the ground, the top bar end and the bottom of the end bar make contact. The part leaning against the hive is the end of the top bar only. These first moves allow beekeeper and bees a calm entry into the hive. The continuing inspection is accomplished with ease in a tranquil atmosphere.

Using the hive tool, move frame number two toward you—so that it occupies frame number one's former position. Each frame is thus moved over into the next open spot. If you remove a frame for inspection, replace it in exactly the same front-to-back and side-for-side position as it came out. In this way you will not crush any bees when all the frames are again in place. One other hint regarding replacement of frames. Always return the frame snugly against the frame previously inspected. No space is left between frames for any bees to be crushed when the frames are moved later as a unit. *Move subsequent frames to adjoining empty slot*

You have finally reached the frame which is holding the queen cage between it and the next frame. Look down through the hole where the candy was located. You should see clearly into the cage. There will probably be a few worker bees in it fussing about. Great! Your queen has been released and you are ready to remove the cage. But you must do a couple of things first. Get the smoker and puff, puff, puff the top bars again. Blow your smoke with the breeze not against it, and with the frames not across them. Before you use your smoker, this time and every time in the future, look to see if the bees are lining up along the top bar, heads up, bodies down in the hive, as though getting ready for a race. This is always a sure signal that a little more smoke is required. *Examine queen cage and use more smoke*

After smoking, examine the situation. You will

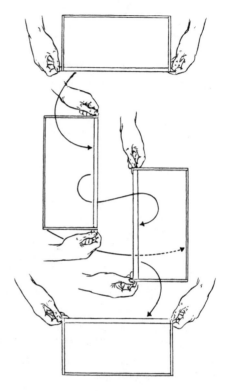

Proper method of turning frame around for inspection of opposite side

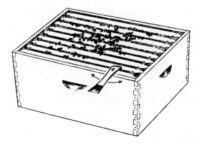

Proper use of hive tool before removing each frame

find that your industrious bees have not wasted the large space made between frames by the hanging queen cage. Secured to the bottom of the cage will be a lovely white natural comb. The bees will bridge it to the adjacent comb or foundation for stability. This extra comb *must* be removed from the hive. Try not to weaken and leave it in the hive. You will regret doing so a thousand times in the future. When bees are permitted to build "free comb" in the hive, it turns into a nightmare for the beekeeper. It becomes almost impossible to find the queen, or even to inspect the comb for any reason.

*Remove any extra comb connected to queen cage*

CAUTION: Use extreme care as you go about the removal of this wild comb. The queen is likely to be close by and it is quite easy to kill the queen accidentally. Before separating the frames holding the queen cage, sever all connecting bridge comb with the straight sharp edge of the hive tool. Then pry off one end of the two frames touching the queen cage. Hold the top of the cage while you are prying. When the cage is free and comes up, so will the wild white comb. If any bees cling to this comb, remove them by "shaking." "Shaking" is a routine mechanical movement you will employ frequently. It allows quick removal of bees from any flat surface: frames, covers, excluders, etc. Raise the cage (with the comb attached) high in the air; bring it downward swiftly, coming to an abrupt halt over and close to the top of the hive. The bees will slide off and into the hive. Frequently, a heavily populated frame may require more than one shake.

*Remove queen cage and "shake" off clinging bees*

I have made a practice of examining a frame before shaking, ever since I injured a beautiful queen in the shaking process. If she is on board, don't

*But don't "shake" off queen*

shake. Remove her first. Queen handling will come to you soon enough anyway. If you have to remove her from the wild comb, just wet your thumb and index finger and grasp her by the wings. Save the comb to examine later. It will probably have a number of eggs in the cells and you'll want to study them at your leisure. Your son or daughter will probably get a commendation if he takes it to class. It is a welcome event with most teachers.

*Raise a frame for inspection of cells for eggs*

The next step is the removal of a frame, if weather permits, for a few seconds of inspection. Examine the cells for the presence of eggs. If you do see a few eggs, replace the frame immediately. Don't take the time to study it on this visit. Be satisfied that you have a laying queen. If you wear glasses for reading, please use them at this time. Without them you cannot see all the things you must. Since I wear them I know you can't see eggs or tiny larvae without them. Incidentally, use of an elastic eyeglass holder will prevent your glasses from sliding off your nose when you bend over. When the frame is replaced, be sure it is snugged up to the one previously examined.

You have completed your first, though brief, examination of your colony. Please do not attempt to do more on this first visit. The weather probably is still cool (late April) and the bees are really not settled in as well as you think. Let's close up the hive and inspect again next week.

*Replace frames snugly*

Place the curved end of the hive tool downward, between the side of the hive and the frame nearest to it. When you swing the hive tool, as you did before to separate frames, all the frames adjacent to each other will move across the hive body as a unit. Do the same action at the other end of the same frame and all the frames will be moved a couple of inches toward the open space left by the queen cage. This works only if we adhere to the rule of replacing frames tightly against each other. Bee losses are reduced to a minimum, and, importantly, so are stings.

As the frames join in the center of the hive, see that the bees are clear of the contact point between frames. Sometimes a moment or two of patience is required before closing the final fraction of an inch. Watch for your opportunity and then shove the frames together. Now space is provided for that tenth frame, left out because of the queen cage. Push all frames tightly together, leaving an equal space between the side of the hive and both outer frames. *Add the tenth frame*

Pick up the inner cover and tap the end on a nearby rock. If it is given one good sharp rap, all the bees on it will slide off and you'll squeeze nary a one as it is replaced. Slide the cover on, starting at one corner, and again you'll minimize damage to the bees. Replace the outer cover in a similar manner, clearing it of bees, if required. Normally you will push it forward to keep the ventilation hole open; but at this point, the end of the first week, you will be covering the empty hive body surrounding the jar of syrup covering the hole in the flat section of the inner cover. *Replace inner and outer covers*

## Some Abnormal Situations

Well, it seems that everything went well and all was normal. However, it might not be that way. *A missing queen*

Sometimes you cannot find evidence of a queen's presence. Evidence of a queen can always be ascertained when you find eggs. No eggs usually means no queen. To be sure, then you must find the queen. This is usually difficult for a novice beekeeper—not because a queen is difficult to find, but because anxiety and nervousness interfere with the concentration necessary to carry out the search.

If you run the misfortune of having a queenless hive, treat the hive normally while you arrange to purchase a new queen. Your packaged-bee supplier is your best source. I shall discuss queen introduction in a later chapter. There are about as many ways as hairs on one's head. One could write a book about that subject alone.

*What to do if there is no activity the first few days*

Another problem that may confront you is a hive of dead bees. It does happen once in a great while. Hopefully, you will recognize the situation as it occurs. If you have observed the colony during the week after hiving, and bees are flying, worry not about this situation. If there was absolutely no activity two days after hiving day, you should make an exploratory emergency inspection regardless of my five-day minimum waiting period. Most often the condition is caused by hunger. Perhaps the holes in the feeder covers are not functioning; or perhaps it is too darn cold for the bees to break cluster. Spray them with warm syrup and you might save your colony. Then fix the holes in the jar's cover.

*Righting a tipped-over hive*

Unfavorable circumstances beyond the loss of queen or bees are very rare. Occasionally a hive is pushed over by an offensive person or animal. If this should happen, it is less of a calamity than one might think. The bees have the talent to adjust to almost every possible situation. All that remains for you to accomplish is the righting of the hive. Straighten it up one section at a time, not all at once. Work slowly and carefully. The bees will tolerate you and there will be no fuss.

# 8

## The Second and Third Weeks

If your colony is developing normally, you need not bother your bees again until the end of the second week, when you should make sure they have plenty of syrup and, if the weather is right, inspect the frames for evidence of egg production.

*Prepare to open the hive as before*

Even though this is your second look into the hive, you will find this visit full of first-time experiences. Fire up the smoker a half hour prior to the opening of the hive, don your protective coverings, and follow the same routine as before in approaching and opening your hive. Don't forget to remove the first frame in order to ease manipulation of the rest.

*Raise a frame with expanded comb and inspect the cells*

Move toward the center of the hive, frame by frame, until you reach one that appears to have foundation expanded into comb. Hold it by the ends and raise it slowly out of the hive. Turn your body to the right or left, while holding the frame, until the sun is over one of your shoulders. Use your third (middle) finger as a support behind the frame's end bar, to angle it into position for the sun's rays. Now you can see into the cells. If it's a breezy day, try to put your body in the way, to protect the frame. Now we are ready to inspect! Inspect for what? Most beekeepers ask themselves that question frequently, but never answer it. Set up a plan for each inspection and then it becomes purposeful, not just a random "Well, let's see what's going on."

Nearly every time you open a beehive, your main search must include looking for eggs, larvae, and any signs of disease. Along the way you will assess the queen's performance by the appearance of the brood pattern. In the early stages of the colony's life we must concern ourselves with the development of comb from foundation. Later you will evaluate the colony's production of honey and pollen. Early guesstimates of honey production can be made by noting the number of cells that are wet with nectar. Finally, there is always a search for queen cells, that sometimes ominous signal of your colony's intention to swarm.

*Judging the bees performance*

Since there are so many details to check over during an inspection, I have found myself organizing each visit by assigning priorities. The timing of the visit usually establishes the priority and then I plan the detail. On your first look into the hive last week, your only concern was with being certain there was a queen and that she had been released. Today you must look for positive evidence that the queen is functioning properly. That means you examine the comb carefully, judging how many eggs she has deposited in this first full week of duty. I don't mean for you to count the eggs; estimate the count by frames. If two sides of one frame are two-thirds full of eggs and/or larvae, your queen and your bees are doing well. If only one side of one frame is filled, they can only be rated fair. Less than that is poor. If the latter, wait it out one more week to see if there is improvement; if the situation has

*Estimate the eggs per frame for good or poor performance*

not improved by then, you should consider replacement of that queen.

*Look for queen cells*

The bees might well be considering it before you! Look for a peanut-shell-shaped appendage hanging on the face of the upper two-thirds of a drawn comb. This is a queen cell, but a special one called "a supersedure cell." If the bees are not satisfied with their queen, they will build this structure and place an egg in it from another cell. Sometimes the queen deposits an egg in it. Then they feed this egg huge amounts of royal jelly. Sixteen days later a virgin queen emerges. This new queen (daughter of the original) is fertilized by a drone, usually within three or four days of birth. When she returns to the hive after mating, she generally disposes of the queen mother. Occasionally a virgin queen will kill the existent queen before mating. There are some cases of mother and daughter living in the same hive for a while, but they are rare.

*Why a superseding queen slows honey production*

All in all, this supersedure (replacement of the queen) has taken about three weeks. This is a long period of time when you consider that the average full length of a season for honey production is a little less than four months, about fifteen weeks. That means three weeks' loss of production is roughly 20 percent of the season; a week is 7 percent. It is far better for you to introduce a new queen obtained from a local supplier. Even a phone call to a bee farm in the South will bring a mated queen within three days by airmail, live delivery guaranteed. Three days more for introduction and you've cut your lost time by two-thirds.

*Variations in queen egg production*

Let's look at the good side and see how the calculation of satisfactory queen efficiency comes about. One side of one frame has about 3,500 cells

(55.3 cells per square inch, both sides of a comb). If a queen deposits about 1,000 eggs a day (about average) for the seven days of the second week, and if she is a good queen, she will fill one frame on both sides. If she is inefficient she might spread the eggs over two frames, filling only a portion of each of the four sides. Naturally there will be a variation from this average figure, both up and down. Some queens can lay almost 2,000 eggs a day; that's about eighty eggs an hour, every hour; or better than one a minute, every minute.

Then one must consider that her majesty needs the cells to deposit these eggs and that's up to the bees— and you. If nectar, in the form of sugar syrup, is provided, the bees will make the wax and draw the comb. That's why in middle April we must help them by feeding, since a plethora of nectar is not available.

Are you tired of holding that frame of bees in the sunlight? That's where I left you after removing the first frame of *comb*. The chances are very good that you have had your arms extended. Let your arms relax until your elbows and upper arms point downward from your shoulders. If your anatomy permits your elbows to rest against the forward side of your rib cage, then you are one of the lucky ones. It is the most comfortable position while studying a frame. This does bring up a very pertinent detail, however. When a frame is close to your face, refrain from breathing upon it. Try not to cough or sneeze on the bees. They don't seem to take kindly to our exhalations. Perhaps they can smell the (odorless) carbon dioxide.

*The best way of holding a frame*

If you see eggs and larvae in abundance, then the

*Noting the
queen's
efficiency
from the
pattern of
egg distri-
bution*

queen's production requirement is adequate, at least for the moment; but how about rating for pattern and distribution? How many cells were skipped in the total area of egg laying? If she misses a cell here and there, it is of no consequence; but if the queen has overlooked enough cells so that a salt-and-pepper pattern appears, her efficiency must be questioned. Include larvae in your survey picture, since the eggs will hatch into larvae on the third day.

*The
advantage
of ordering
a clipped
and marked
queen*

It is not essential actually to see the queen at this time. Once the eggs are visible, we are sure we have a queen; at least we had two days ago. There is an advantage, however, in observing the queen in early weeks, since later on it will prove difficult with many more bees in the colony. If you ordered your package of bees with your queen "clipped and marked," which was an option open to you, then the task of locating the queen is made much easier. The shiny black spot behind her head, which is her thorax, will have a touch of colored paint applied to it. It is harmless to the queen and helpful to you. Besides simplifying queen location for you, you'll always know if you have the same queen or if she is replaced through supersedure. You also will have the advantage of knowing her age, in case you will decide on queen replacement after a certain period of having the same queen.

The "clipping" refers to a procedure widely used to prevent losing the queen. One of her wings is cut back somewhat, which makes her unable to fly. This is not a cruel procedure as she need not fly except on two occasions. The first time is when she has her nuptial flight. This is accomplished before you get her. The second time is when she is leaving your hive with a complement (usually about half the colony) of bees to find another home. This is of course, swarming, which you will do everything in your power to avoid (see page 103).

One of the first things you can do to sidestep swarming is have a "clipped queen." When the bees issue forth from the hive, they may fly to the nearest tree, forming a large cluster. This is their first step while waiting for scouts, who are searching for a new abode, to return. In the meantime, the queen, who tried to leave with them, finds it impossible to fly and crawls back to the hive. When the bees find she is not present in their cluster, they return home.

## End of The Third Week

By the end of the third week things are really beginning to look up. It really becomes quite simple to evaluate the queen's total competence by this time. On the ninth day, the larva enters the final or pupal stage. This is the time when the bees enclose the cell with a flat capping of beeswax. The color ranges from light tan to dark brown. The first eggs will reach this stage of development during the third week after you have established the colony. These large unbroken areas of capping (capped brood), display the queen's expertise in egg laying and distribution. (If she can be judged at the end of the second week, so much the better, should there be some failure.) After you have opened the hive, work toward a capped frame and start judging. Usually there are no eggs on such a frame; therefore, in order to check for the queen's presence, you will examine frames just before or just after those that are capped.

By now enough time has passed since hiving your

*Judging the queen's performance from capped brood cells*

bees to check regularly for any evidence of disease. What you look closely for is a salt-and-pepper pattern, very much like a poor queen's egg-laying pattern. It is much the same: the salt-and-pepper appearance of brood here and there. The major difference between a poor pattern and disease can easily be seen upon closer examination. The cappings are sunken, concave rather than convex, and usually have perforations in their centers. This is the first sign of trouble. The healthy cappings are generally slightly raised, or at least flat, but never sunken or perforated. The larvae (uncapped) should be white, almost glistening, when healthy.

European (EFB) and American (AFB) foul brood are the two major diseases mentioned earlier. They are both deadly and can cause the death of your colony. As the name suggests, the brood is attacked. If no bees are born the colony will die as the adult bees live out their life span and die. The adult bees are unaffected by these two diseases. AFB is the more deadly of the two, since it has a spore stage which causes a continued reinfection. It is extremely contagious and must be checked promptly. It is up to you to discover either of the foul brood diseases in their earliest stages, using these known signs and symptoms of trouble. Then get in touch with an expert beekeeper for positive identification and treatment. Better than that, contact your state bee inspector, who is truly qualified and will help you. Remember, sunken, perforated cappings over the brood, scattered here and there on the comb, should alert you to the problem.

The end of the third week is a good time to adjust those frames of foundation which are not yet drawn into comb. As you reassemble the hive after inspection, you may take a frame that the bees have not reached and place it between frames of drawn comb. Do not place such a frame between combs of eggs or larvae. The rule is that you may always put frames up to, but never into brood frames. The bees organize their brood nest in a compact manner that must never be disturbed. Later on, in a fully established colony, some of these rules can be broken; but during this early critical period treat them with great care.

During the next three weeks you will want to develop your ability to identify the different materials brought in by the bees. Pollen comes in as many different colors as there are flowers. The cells near the larvae will begin to fill as the supply of this protein source increases. There will also be a noticeable increase in the number of cells that have a "wetness." It could be nectar, but might also be water. Generally the nectar is stored in the upper corner areas of the frame, referred to as the "upper quadrants." The brood frames begin to take on an almost patternable appearance, which you will learn to recognize as a "good frame" with a quick sweep of your eyes. The brood occupies the central area with a band of pollen around it and honey above. The two lower quadrants will have brood clear to the frame, from a prolific queen. If the area is not filled with eggs, larvae, and finally capped brood, the bees will use it for pollen storage. Pollen is generally not capped over directly, but later in the year, in preparation for winter, the bees will cover it with a little honey and then cap it. Honey capping is light in color, sometimes pure white. Brood cappings have colors ranging from light tan to dark brown. You can tell what the cells contain by the colors.

*Noting the patterns of cell storage: brood, pollen, honey*

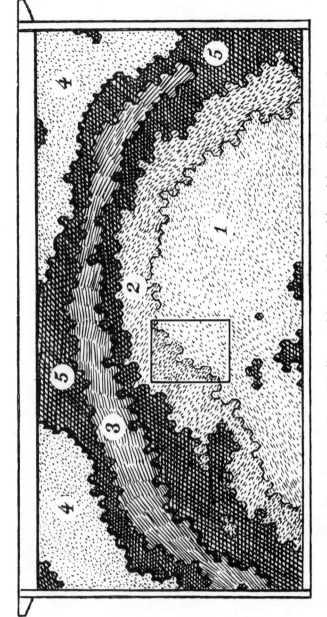

An excellent pattern of brood, pollen, and honey. 1. Capped brood (pupal). 2. Eggs and larvae (hatched eggs). 3. Pollen. 4. Ripened and cured honey, capped over. 5. "Open" cells.

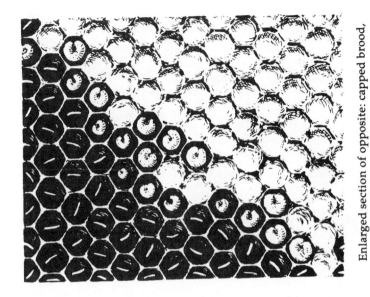

Enlarged section of opposite: capped brood, eggs, larvae

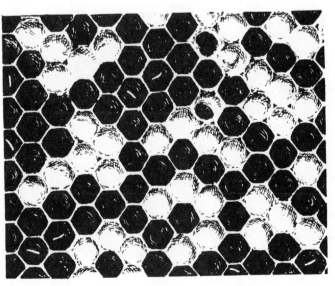

A poor pattern. Replace the queen.

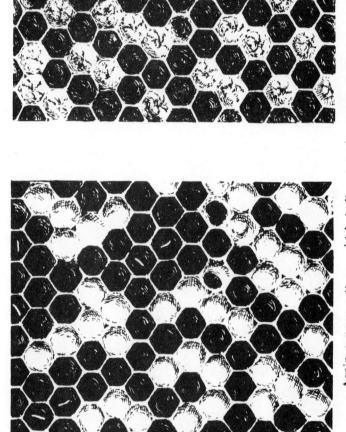

Again a poor pattern, which indicates replacement of the queen—but note how closely it resembles the diseased-type pattern at right, which shows concave cell cappings with perforations.

The best view of diseased larval remains in the bottom of the cells

# 9

## The Fourth, Fifth, and Sixth Weeks

If the weather has been cooperative and you have a fairly good queen, when you visit the hive at the end of the fourth week you will find the bees will have made good progress during the month since coming to live with you. I know you have continued feeding syrup because all the frames are not yet drawn comb, and until they are comb, you'll want to furnish the raw material for the wax manufacturing in a steady supply.

It is probable that by the end of the fourth week six of the ten frames of foundation have been drawn out into comb. Not until the seventh frame is completely drawn into comb will the syrup jar be replaced with ten frames of foundation in the upper hive body. At that time the inner and outer covers will be moved upward as a unit. Even though this hive body will still be used for brood during the first part of the summer, it will eventually turn into a food chamber and contain the main source of the bees' stores for winter consumption.

The proper time to add the food chamber is rather critical. The bees will need the additional space for expansion. The population is increasing now in large numbers every day. The increased army of worker bees helps to expand the total potential of the entire colony. The more bees that are available to incubate the brood, the greater the queen's productivity. More bees available for cell cleaning and feeding the larvae also increase the queen's activity. It becomes a population explosion. You might wonder why the second hive body isn't added earlier, or for that matter, why not start with it immediately when originally setting up the hive? There are several reasons, but the most important of them relates to the bees' manner of nest building. They like a nice tight, compact nest. Their capability of maintaining the necessary incubation temperatures is greatly enhanced by a small enclosed area. In fact, practicing beekeepers from Europe provide help for the bees in their quest for a compact enclosure. They will hive the bees on a half-dozen frames enclosed by two "follower" boards in the brood chamber. A follower board is a piece of wood one-half inch thick. It hangs outside but next to the first and last frames. As the bees develop comb and fill the cells, the beekeeper adds one or two frames at a time. I have tried this system and it does help the bees. You must decide whether you can devote the time required to stay just ahead of the bees when using follower boards.

*Why it isn't added sooner*

I've told you about this because I'm anxious to convey the idea of staying "just ahead" of the bees. If the second deep body goes on too early, you'll dissipate the heat generated by the nurse bees for incubation. In order to generate more heat they'll use

*Staying just ahead of the bees*

up more syrup or nectar, which means less material from which to make wax. That means less comb building, which in turn will affect the queen's egg production and subsequent bee population. Fewer bees means less honey. A well-meaning move can seriously affect the bottom line—production of honey.

*Too much crowding causes swarming*
There is another side to this space allocation in beekeeping. It requires careful scrutiny by the beekeeper to see and understand everything that is happening in the combs. If the cells that are available for use by the bees are filled with nectar being ripened into honey, the honey itself being stored and capped, plus pollen packed for the brood, and finally the brood in various stages of development—what can the bees do when more of everything continues to flow in and space is not provided for them? They do what has been natural and instinctive for their survival over eons of time—they swarm. That is when the colony divides and half of the bees leave for a new home.

The beekeeper is left with the other half, and his opportunity for a rich harvest of honey is jeopardized. If it is early in the season, a colony may be able to recover and store enough honey for the winter. If not, the beekeeper will assist by feeding the bees through September and October in most temperate areas of the country. (November-December feeding is in case of emergencies.)

The situation for crowding and swarming rarely happens at the end of the first four or five weeks, but neither is this dilemma too far into the future. If the beekeeper keeps in touch with his colony, the crisis can be avoided.

Before leaving the hive on this visit, it is wise to move foundation frames inward. Usually the outside frames' foundations are untouched by the bees in the first four weeks. The bees' progress will be helped if the position of these outermost frames is reversed with two frames containing comb. Do not move any frames containing brood in any stage of development at this time.

*Switch outside frames with inner frames*

By the end of the fifth week each frame's increase in weight is dramatic. Frames that were partially filled are now full, containing honey, pollen, and bees. Hold the frame securely and enjoy the bees' performance right there in front of you. Give yourself a treat and watch a fledgling bee chew her way out of the cell. If the temperature is in the seventies and the sun is shining on the frame, give yourself several minutes of studied observance. Watch the capping intently for any movement under it. Sometimes a hairline crack over a cell is a telltale sign of a baby bee's movement. Her mandibles chew out a circular hole through which she will emerge. Her body is covered with a downiness so typical of most newborns. The baby bee's eyes sometimes appear to have a light film over them. It disappears quickly as she cleans herself for the first time. Her activities seem to begin as soon as her wobbly legs lose their unsteadiness. An interesting attitude to the newborn is displayed by the other bees in the hive. At first it appears that they pay these babies little or no attention. It is only when you watch closely that you will see a momentary pause as a bee passes a young one and then quickly moves away. But not every time is the departure so rapid. I have seen an elder, singularly or with others, groom the young one so that she

*Watching the newborn bees*

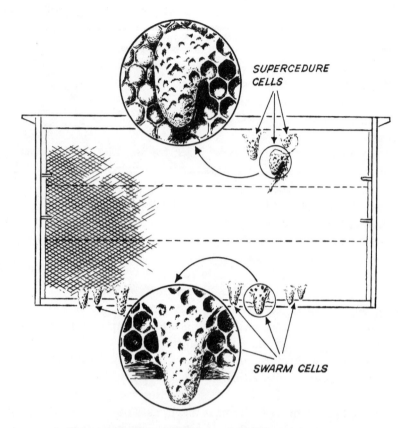

SUPERCEDURE
CELLS

SWARM CELLS

Queen cells—supercedure on upper half of frames and
swarming cells on bottom area of frames

glistens. Why some and not others? I do not know. It
is so pleasant to be privileged to this scene; don't
pass it by.

*Searching*
*for queen*
*cells*

After you have watched the interplay between the
bees, look over the bottom bars and lower area of the
frame. The time has come to search out queen cells.
When built on the lower third of the frame, they

signify the bees' intent to swarm. If there are seven or eight cells present, you may be sure of your diagnosis. Remember, supersedure cells are the same in appearance but there are usually only two or three and they are located on the upper two-thirds of the comb.

The chance of swarm cells in a new colony of this age is quite remote, but you must check it out. There is always a chance it could happen. The month is May and that is the big swarm month. There is a saying which is worthwhile passing on to you here. It will stand you well in the future. I don't know who said it or where it came from—but it goes like this. "A swarm in May is worth a ton of hay. A swarm in June is worth a silver spoon. A swarm in July ain't worth a fly."

*Be alert early against potential swarming*

Not only are the swarms usually smaller after May, but the time grows shorter for the buildup before winter. In the next chapter we will discuss swarming in some detail. It is mentioned here because you must stay alert for the signs of it, so you can deal with it promptly . . . and "It is May," the bees seem to tell each other.

Throughout these five weeks you have maintained a very small entrance opening for the bees. The boardman feeder (entrance feeder) has been used along with the inside feeding. Now the bees have established themselves and the way is clear to enlarge the entrance. If invaders do try to enter the hive and if the colony can't handle them, then you will restrict the entrance again for a temporary period. (How to cope with robbing bees is explained on pages 124–126.)

*Opening the entrance*

Remove the entrance feeder and take the entrance

*Use boardman feeder and grass as opening reducer*

reducer away completely. Replace the entrance feeder into the hive entrance, but tightly against one side of the hive. Stuff the entire entrance with grass, from the feeder over toward the opposite side, leaving an opening the width of about four fingers on the side of the hive opposite the feeder. The grass should be tight enough to restrict any bees from passing through it. Each week after this take away another inch or so of grass until it is gone. Continued use of the boardman feeder will insure the bees a nectar flow to sustain brood rearing and wax making.

*Continue syrup until bees stop feeding*

When a real nectar flow from the flowers becomes available, the bees will stop taking syrup from both inside and outside feeders. The bees will tell you when they no longer need your support in feeding. That moment usually arrives at the same time as warmer nights (above sixty degrees), and the entrance may be left wide open. If it turns out that feeding is no longer necessary, but the nights remain cool, then reinsert the entrance reducer, using the largest opening. This will prevent chilling the brood.

The inverted gallon jar has been the main source of syrup for the bees. It has been available day and night, in clear sunny weather as well as the inclement periods. The boardman feeder was used by the bees only when warmed by the temperature on milder days. Continue feeding with the inside container until either of two conditions occurs. The bees will stop taking syrup as soon as sufficient nectar is available from the flowers; and that means the feeder is no longer required. Or as soon as the second hive body is required, feeding may be discontinued.

By the end of the sixth week, forty-two days have elapsed since the bees were hived—enough time for two generations of bees to be produced, since gestation covers a three-week period. Not really though, since the queen's release did account for a few days in the beginning. Startling enough, however, is the result of an elaborate calculation revealing forty thousand bees in larval or pupal stages and at least fifteen thousand bees hatched out as adults.

If the second hive body has not yet been added, the chances are that it will need to be added now. Inspect the frames and make an estimate of the number of bees that could be hatched out in the next week or ten days. If it seems that three or more frames of brood will hatch out, the second brood chamber, complete with ten frames, is now required.

*Adding the second hive body*

At this point progress can be continued through the use of a simple manipulation, which is simply the intermingling of new frames in the second hive body with brood-bearing frames in the original hive body so that both bodies contain new and full frames. Remove five frames from the new brood chamber, but keep them handy.

Here is how to do it. Smoke your bees in the usual manner and remove the two covers and the outside frame in the lower brood chamber. Place the second hive on the reversed outer cover sitting flat on the ground. Divide the remaining five frames in this hive body—three on one side and two against the opposite side wall. Now move into the brood chamber and remove four frames of bees from the center, one at a time. Place these frames in the center of the new hive body in exactly the same position and relationship to each other that they occupied in the original brood chamber. Take note of the frames' contents as the transfer is made. There should be at

*Intermingling old and new frames for best results*

least two frames of capped brood (some open larvae are all right) containing some honey and pollen. The other two frames may be the same, but more likely will have more pollen and honey than those two in the dead center. If some adjustment is required in order to have some honey and pollen on the outside two frames, then make it. These frames are transferred with attendant bees. Take great care in handling so that no bees are crushed. The queen might be one of them. Now push the foundation frames toward the brood frames until there is room left to add one of the five frames you removed earlier. Move over to the main brood chamber and *carefully* bring all the frames remaining in it toward the center. That brood chamber is short four frames, which will be added by placing two foundation frames on each side. Place the new hive body above the brood chamber, carefully smoking the bees in the lower chamber to drive them down, thereby minimizing the danger of crushing them. Again, one of them might be the queen. It is surprising how thoughtful and gentle one becomes when the queen is kept in mind.

In nature, the bees including the queen prefer to build the brood nest upward in an elliptical or teardrop shape. Moving the four brood frames upward will serve that purpose and also a few others. The heat from below will help those above and actually fewer bees will be required to incubate the same amount of brood. Then too, these upper brood frames act as "bait" for a more rapid expansion into the upper chamber. Finally, you have taken the proper step forward by providing space at exactly the moment when it will be required.

# 10

## Swarming

The owner of a first-year colony will probably not run into the problem of swarming unless his queen is very prolific and the early nectar flow has been intense. However, swarming creates a problem for any beekeeper, and as it sometimes can happen to the beginning beekeeper after the eighth week, the subject needs to be covered here.

Nothing upsets the activity and production of a colony as much as swarming. Swarming greatly reduces your chances of harvesting surplus honey, since the foraging force is cut in half, sometimes more. Usually the bees cannot build up their population and stores of honey, in the remaining time left in the season, to provide a surplus. In fact they might not even get enough together to take them through to the next season. You will probably have to supplement their stores for winter. Apiculturists all over the world have devised and tried systems of every kind; but bees go on swarming. Their natural instinct for survival spurs them on to increase their population until it spills over. Then they divide into two colonies and start anew. This has been their

*The harmful effects of swarming*

system for thousands of years.

The attentive beekeeper can establish a proper environment for the colony and minimize their itinerant behavior; but the novice beekeeper nearly always seeks out a veteran bee master who offers some time-honored system, rarely effective. Instead, I propose that we examine closely those conditions most conducive for bees to swarm and see if we can prevent them.

The predominant cause driving the multitudes from their home is congestion in the hive. Crowding through overpopulation is the stimulus that triggers the instinct to leave the hive. Once the provocation has taken hold, it appears that their minds are made up and flight is irreversible. The story goes, "Once the bees have taken it into their heads to swarm, there's nothing you can do." Nonsense! There are all kinds of incentives to change their objective; but, before I describe these procedures, let us look further into this business of "overcrowding," also referred to as "congestion in the hive."

It should be obvious by now that a colony requires sufficient space to carry out all the steps used in the honey-making process. Even with the bees' masterful use of space, they do run out of it. They cannot store pollen and ripen honey and rear brood in the same cubic area. So, as the weeks of late April and early May fly by, the beekeeper must be aware of his colony's requirements for more room *on a timely basis*, not *after* the space is needed.

I listen to a frequent lament by novice beekeepers who insist that additional space was provided for expansion. "I gave them a second deep hive body full of frames and foundation and still the bees swarmed," states the beekeeper.

Since the hive body was empty and unused when the bees left, the beekeeper is rightfully indignant that his offering of space was ignored. From the bees' point of view, the space came too late. All the nectar was hanging out to dry, pollen filled a large area, and bees were popping out of cells by the thousands. What could the colony possibly do when a super full of frames with flat sheets of foundation was placed above their brood nest? There wasn't time to expand those sheets of wax into comb so the cells could be used. There was just nowhere to go but out.

*The importance of anticipating the need for more space*

The act of adding another hive body is always correct; it is the timing of the act that is usually wrong. It is usually done too late. Picking the proper moment to add the space is difficult, but certainly not impossible, to judge. I have always used a 70 percent rule of thumb. If ten frames are in a hive body, then seven frames are 70 percent. When this many frames of foundation are made into drawn comb, give the bees more space, regardless of what may be in the cells. Three frames are left for the bees to work on, while those above can be tempting if a few frames of brood are raised up to bait them.

*The 70 percent rule of thumb*

Ventilation, or rather the lack of it, is also a contributing cause of swarming. It is so closely connected with the congestion problem that it is frequently overlooked. A strong probability exists that, when a colony is crowded and swarms, they might have remained in the hive if ventilation had been adequate. Instead it added just the necessary discomfort to turn the bees in a definite direction.

*Ventilation and swarming*

Hot weather in early spring usually forecasts a strong swarm period ahead. If a colony is located in

*Opening
the colony
in hot
weather*

bright sun and no shade is available to them naturally, it must be provided. When it is required, one may
stagger the individual hive bodies to assist the escape
of heat and vapors. If a good nectar flow is on, there
is no risk of creating a robbing condition. The bees
have a fine supply of nectar from the flowers, and
they are far too involved in foraging to consider
anything else. If the colony consists of two deep hive
bodies, the upper one may be moved forward slightly to create an opening in the top rear of the lower
body. At the same time the overhang in the front by
the upper body protects the bees below from inclement weather. If the opening between the two bodies
in the rear is too large, a wooden entrance reducer
placed on the ledge formed by the top of the lower
body will give protection from weather there. The
bees may be permitted entry and egress through
these new hive openings for the period.

The half-moon hole in the ledge of the inner cover
must remain open, as well as the oval hole in the flat
section of the cover. Be certain that the overhang in
the front of the outer cover does not block the flow
of air. Pushing the cover forward, when standing to
the rear of the hive, will assure the proper position.
Elevate the inner cover one-sixteenth inch above the
hive body by providing each corner with a wedge of
that size (see page 15). Secure the wedges to the
inner cover, for they help provide ventilation in the
winter as well.

*Have water
nearby*

Provide the bees with a source of water nearby so
they can cool the hive themselves; but save them a
long trip to get it. A water-carrying bee might double
as a scout bee and tell the colony members about a
pleasantly cool fir tree that's just right for a new

home. Then off they would go, all for a drop of water.

It is actually a collection of little things that compound to bring about a swarming situation within the hive. The large considerations which induce swarming, however, are lack of sufficient space for the bees to accomplish all functions and a dense, overhumid, piping-hot hive.

Sometimes the beekeeper recognizes the symptoms of an imminent swarm in spite of furnishing the bees with all the proper conditions to stay home. When this happens, you might still prevent the swarm from leaving the hive by using some radical alternatives. Generally, by the time a coming swarm is obvious enough to diagnose, and it is really going to happen, then strong methods must be employed in order not to lose the bees. Even having a queen with clipped wings who crawls back to the hive when she cannot fly—as mentioned in an earlier chapter—will only delay a second swarming attempt about eight days. This is the time it takes for a virgin queen to be hatched out of one of the many queen cells (swarm cells in this case) that were set up by the bees to insure the survival of the parent colony. Before I describe some measures which can be of help, let's take a look at some of the bees' *requirements* to leave in a swarm. These are different from the conditions that *cause* a swarm.

*Recognizing the bees' requirements*

The colony prepares to divide and leave for another home only after they are certain there are sufficient bees and *brood* left behind to sustain the parent colony. Of course, this is usually the case, since overcrowding wouldn't be a problem if there were empty brood frames. This gives the beekeeper a

*Bees will not leave a hive that has no brood*

number-one control method. If all the capped brood frames are removed from the hive and replaced with frames of foundation or drawn comb, preferably the latter, the bees will not leave. I'm sure it appears too simple, but under normal swarming conditions the bees will not leave a hive that has no brood. There are several ways of going about this.

*Therefore, you can divide the parent hive, gain a new hive, and prevent a swarm*

The most common way to create the situation of an old hive with no brood is to place a new hive body a few feet away from the parent hive, move the brood frames and attendant bees from the parent hive into the new hive, place the new frames in the center of the new hive and surround them with additional empty frames. If some eggs are present on one or more of the frames of brood, the bees will make a new queen for themselves. You might want to speed things up and supply a purchased queen. The foraging bees that are transferred with the frames from the parent hive to the new hive will rejoin their parent colony, since that is their home. Even though all the brood has now been removed from the parent hive, the colony is heavily populated with adult bees and the queen now has new cells to fill with eggs. Three weeks from this time, new bees will again appear. In the meantime, the new colony's nurse bees will be pressed into service and begin foraging. If a frame or two of pollen and honey are supplied as well, it will make the growth of the new colony a certainty. Be certain the queen is not present on any frames used to set up the new colony.

*Two hives are better than one*

Actually, having two hives of bees is easier than having only one. You can "borrow" from one hive to help another; if one is weak in the spring, the strong hive can supply a few brood frames, or honey, or pollen. If one hive goes queenless, eggs can be given to the queenless hive to make a queen. Also helpful in the fall of the year, a prosperous hive might have extra frames of honey that its weak relative might need desperately to carry it through the winter.

However, if an increase in colony numbers is not desired, it is easy enough to combine the new hive with the parent colony; but only after the swarming urge has passed. Removing the brood and permitting it to hatch out provides the parent colony with two benefits. First, they have the needed room immediately with the new frames that replaced those occupied by the brood. Second, if you do combine them into one colony but wait twelve days to allow the brood in the new colony to hatch, the comb will have empty cells, which the parent colony will be able to use immediately.

*Combining two hives*

Joining two colonies can almost always be accomplished successfully using the newspaper method. The bees must chew their way through a single sheet of newspaper which separates them. As they perform this act, the hive odors mingle and all is peaceful when they unite. A single sheet of newspaper is spread over the original hive's top deep chamber. The new hive is placed on top of the newspaper and the inner and outer covers replaced. Use a small-diameter nail to make a few holes in the paper. This will help the bees get started as well as provide the upper hive body with some air. The new hive must be examined for queen cells before being placed above the parent colony. If any are present, they must be destroyed. Just twist the cell and remove it.

*The news-paper method*

The second requirement the bees have for swarming is a queen to go with them. Usually the original

*Bees will not leave without a queen; so remove any queen cells to prevent swarming*

mother queen, frequently referred to as "the old queen," will go with the swarm. Normally, about eight to ten days prior to swarming, the bees will make seven to ten queen cells along the *lower* portion of the frame. An egg in one of these cells will develop into a virgin queen, who will be the parent colony's future queen after mating. If these queen cells are methodically checked and destroyed, the bees will not swarm. They will try, however, to build another group of queen cells, which in turn must be destroyed as well. This method of swarm control does leave a great deal to chance, because it is easy to miss a queen cell when checking. That is all they need; from one queen cell comes one queen and away they go! Destroying queen swarm cells is widely practiced as the best method to prevent swarming. It seems to me to be fighting with the bees, instead of joining them in their quest to satisfy their instinct.

*How to harness the swarming instinct*

How can we join the bees and still avoid losing them in a swarm? Well, it has to do with their very own requirement for swarming. Their third requirement is obvious. There must be a sufficient number of bees to split the colony! Suppose we satisfy the bees' urge to swarm, but with enough supervision and control so we don't lose them? In that way we essentially will be joining the bees instead of going against them.

*Creating an artificial natural swarm*

An old bee master taught me how to create an "artificial natural swarm." Make a tripod out of three saplings, about seven feet long. Tie them together at one end, and they'll stand up well enough for this manipulation. Next, move the hive which you think is going to swarm a few feet to one side. Over the spot where it had been set up the tripod. Hang a tree branch from the tripod almost to the ground. Remove the frames from one of the hive bodies one by one and shake the bees on to the hanging branch of the tripod. If you use the frames from the super that had the queen in it, the chance of success is better. The bees will cluster on this branch the same as though they had swarmed. If the queen is with them, they can be hived anywhere you please. Just give them a couple of hours to sit on the branch, then shake them and the queen into a new hive in a new location. The old hive can be moved back to its original position after the tripod is removed. The "swarm" can even be hived back into its old hive, providing the hive is kept at a *new* location. This "new" location can be three or four feet away from where it previously stood.

Well, there it is: the conditions and the causes of swarming, and the essentials the bees say must exist before they'll swarm, regardless of the conditions. A good beekeeper will try not to allow these conditions to develop; but, if they do, a good beekeeper will utilize the bees' requirements as the tools to prevent swarming. If it is believed that the exercise of dividing the hive or creating the artificial swarm is too much work, then reflect upon the difference between swarming and dividing. When a swarm is lost, you lose roughly one-half of the hive; when you divide the hive, you have doubled the colony number.

All too frequently, every tactical measure we execute ends in futility and they swarm anyway.

"Swarm fever" is a term understood only by the beekeeping fraternity. The most placid people become all energy, pent-up dynamos gone wild. An incoming phone call notifying one of a swarm will incite a furiously rapid gathering of swarm-catching gear. The vehicle leaves the garage with squealing rubber and he is on his way to ensnare a swarm of bees, which later will breathe life into an empty beehive in the apiary.

*Catching a wild swarm is not as difficult as it looks*

Catching a wild swarm of bees gives one a real thrill. There is a spirit of the hunt coupled with danger when faced by thirty thousand spear-carrying bees. There is also the reward of bringing them home and starting a booming hive. The promise of honey from a swarm is strong. It is that promise that overcomes the natural fears one has before catching one's first swarm.

It is only after that first swarm is caught and hived that the beekeeper confirms the ease of completing such an operation. There are a few good reasons why capturing a swarm is generally not difficult.

Swarming preparation in the hive includes many activities to insure the swarms' security for several days after leaving the hive. After all, they know that they may hang like a huge teardrop from some tree branch for as long as a week. It might be that long before the scout bees will locate a proper tree, be it a fine hickory or rock maple with just the right-height knothole leading into the heartwood. It is there in the tree's center they will build a beautiful nest brimming with early bright combs that fill with a kaleidoscope of pollens and honey during the months ahead.

At some point, close to their time of departure from the hive, the bees, just as all travelers do, pack their valises. Honey stomachs are filled to the fullest to sustain the colony through the days ahead. Swarms are usually extremely gentle for this and many other reasons. I have never caught a swarm of bees whose disposition has been other than genial. However, I must admit to following the rules when swarm catching! I never try for a swarm if I'm told that the bundle of bees have been hanging there for four or five days; especially if there has been a rain shower or two during that period. I also make it a practice to do my bee hunting during the sunny part of the day and never during inclement weather.

*Swarming bees are usually gentle in sunny, dry weather*

Capturing a swarm is in itself a simple matter, or extremely difficult, always depending upon the location and height of the bees. There are many different methods for getting a high swarm; my own is to limit the height factor to no more than thirty feet. Any swarm above that height will never find their way into any of my bee yards unless they fly there. Thirty feet can be reached by ladders, poles, and sometimes a rock, thrown over the tree limb, carrying a light line with it. Once the line is over the limb and back down to the ground, a heavier rope is attached and drawn over the branch holding the swarm. Now you have many options open to you for hoisting something aloft. A bushel basket, containing a frame of brood, is irresistible to the bees if it can be brought close to them. I have enjoyed lowering such a basket filled with ten pounds of beautiful, docile, honey-filled bees.

*Capturing a high swarm*

If you are not the best rock thrower, use a toy bow and arrow, or even the real things if you own them, to place the original light line over the limb.

It is always more pleasant to catch a swarm if you are sure they are a peaceful bunch, since there are always exceptions. Old-timers never wear bee veils when catching swarms, but I do, and advise everyone else to do the same. Five or six well-placed stings on the face can make an awful mess for a few days. Why bother trying to reach hero status when a proper head veil precludes any such dire consequences occurring during an otherwise fun-filled event? Wearing gloves for protection is an option. I advise against them simply to avoid awkwardness, which is often the direct cause of receiving a bee sting.

Well, if the bees are reachable and you are equipped with head veil, test them out by placing your hand two or three inches from the swarm. Get your hand close enough to feel the heat radiating from the group. If they are a serene group, your hand will be ignored, except perhaps for a curious bee or two which may light upon it. If less than honorable intentions fester within, they will indicate their ill will by attacking your hand, not en masse, but perhaps one or two bees will fly at it like a dart. It is quite different from the inquisitive bees that fly first above it, then settle gently upon your hand. Trust them; for I repeat my first remark about their dispositions when swarming. I have never had a swarm punish me, in spite of some mighty poor handling on my part.

Should the swarm be low enough for your pole to make contact, secure bushel basket to its end and raise it under the swarm. Once it is in position, a sharp jab upward will dislodge the bees, which will fall into the basket. Lower away and either cover the basket or transfer the bees to another box prepared for them. If you have the queen in this original bundle, the bees that are still airborne will gradually find their way to this box with the queen. It is a pretty sight to see fifty or sixty bees, fanning violently, pointing at the box, saying, she's here, she's here; and if she is, the box may be opened and the bees will not leave. In fact, those on the outside will rush to join the crowd inside with their regal monarch.

*Capturing an easy-to-reach swarm*

Hopefully, you will have good luck first time out, and find a beautiful swarm suspended from a branch about shoulder height. If the branch is about thick enough to be severed by lopping shears, then do so gently while holding the branch so it will not fall to the ground. Then take the branch to the swarm box or hive you have with you and place it carefully inside. You certainly have the queen here—for you have the whole swarm.

*Capturing an almost inaccessible swarm*

Sometimes the bees are in an almost inaccessible place that doesn't lend itself to the forms of capture I have described. If any bees are open to your reach, brush as many as you can (with a *soft* brush) into a small peach basket. Try to get a cluster into it at least the size of your fist. Put the basket with the bees about five feet away from the main swarm and, if possible, about four or five feet off the ground. Now fire up your smoker, and blow clouds of smoke, but not fire, on the main swarm. They'll take to the air, and if you're as fortunate as I have been, they'll wind up settling in the basket.

*The advantages of a swarm box*

Mirrors, fires, anise, and beating on pots and pans are all methods of capturing swarms I have never used, which are supposed to work. But I would like you to have the best tool of all for seducing these tiny rovers back to your bee yard. It is a swarm box, designed by a friend of mine with features that

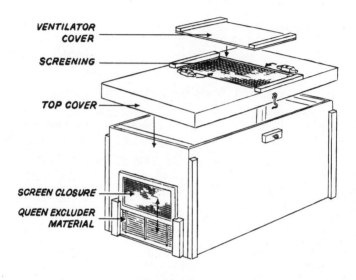

VENTILATOR COVER

SCREENING

TOP COVER

SCREEN CLOSURE

QUEEN EXCLUDER MATERIAL

The swarm box

makes it comfortable for the bees as well as the beekeeper. The accompanying illustration clearly shows the details. The size is not critical, though mine is 21 inches long by 10 inches wide by 14 inches high. I believe a smaller size would be impractical and a box too much larger would be cumbersome and hard to store. Make it of light materials, ¼-inch plywood is sufficient for the sides and ½ inch by ½ inch lumber for the frames, which will keep the weight down.

The swarm box will never leave a single bee behind when a swarm is captured. No other system can make that claim and you'll see why. The basic concept in corralling a swarm starts with the containment of the queen. Wherever she is the bees will follow. With this in mind, my friend decided that the entrance to the swarm box should be constructed of queen-excluder material. This is composed of horizontal or vertical wires which will not permit a bee the size of the queen to pass through, but will admit all worker bees. Small drones will make it through, but most drones will not. The top of the swarm box is completely removable and part of it is screened over to be used for ventilation once the swarm is contained in it.

Whenever a swarm becomes available, set up the swarm box on the ground near the hanging swarm. Now go after the swarm in whatever manner is appropriate to their location and availability. If the swarm ends up on a branch that you sever from the tree, carry it over to the swarm box. Remove the cover completely and then place the branch and bees inside, if the branch is not too large. If the branch won't fit, hold it high in the air over the swarm box and make a fast downward move, stopping abruptly just short of the box. Most of the bees and hopefully the queen will fall into the box. Put the cover on immediately. Do not shake any more bees from the branch after the first time. Raise the screened entrance slide, thereby opening the entrance. The bees inside may come out to see what's going on, but they will turn around and go back in, *if the queen* is inside. All you need to do now is to lay the branch down a few feet in front of the entrance to the box. The bees remaining on the branch will head for the swarm box and their queen.

*Using a swarm box*

I guess you have the idea by now that we need the queen inside the box for it to be effective. Actually it becomes effective even when the queen is not in the first group of bees that were shaken into the box; but then you have to be more attentive. It works this way. After shaking, watch the bees entering the box. If they continue to do so and a number of bees begin to fan with their wings, all the while pointing at the box, she is probably inside. However, if you notice a slackening to the entry movement, look quite carefully around the entrance for a clump of bees. Sometimes, when the queen is not in the box, she can be discovered at the entrance trying to get in, which she cannot do because of the queen excluder entrance. Wet your fingers and pick her up by her wings and settle her into the box. Now sit down and watch the fun. The bees will climb all over each other trying to get inside to join the queen.

*A memorable experience*

I remember one swarm I was after a few summers ago. Something unusual occurred that I can never forget. The bees were clustered on a shrublike tree, about eight feet in the air. They were fairly close to the trunk of the tree, so I had to pull out my pruning shears and work my way in, clipping out branches here and there. I finally reached the swarm and began to saw the thicker branch they had chosen. Since it was an arm's length above my head, it was rather awkward. I began to wonder if I was in trouble when the saw cut through and, though my left hand held the branch, it sprang upward dislodging a great portion of the swarm. As they took to the air, I decided I would try to use that part of the swarm still clinging to the branch. I put it into the box, praying that I had the queen. After closing the box, I stepped back a few feet, watched and waited. The bees seemed to be going in, not too rapidly though, and I started to suspect failure. The bees still outside the box were flying round and round the swarm tree. Some were in and on branches that were close to the stump of the one I had cut off.

I moved the swarm box some thirty feet away to a shady spot, as the day was blistering hot. I was pretty hot too and sat down on a rock to catch my breath. As I watched the entrance to the swarm box, a ball of bees grew larger right in front of me. I knew at once the queen was there, *outside* the box. I put my reading glasses on and went down on my knees. I stiffened my index finger and gently began to probe into that mob of bees. They were kind of churning as they formed the ball, so, as I moved some out of the way, others filled in the hole made by my finger. I kept trying, almost sure that she was there; and suddenly there she was, a beautiful pale golden queen with an abdomen like a canal boat. I picked her up with my right hand and with the other opened the box. In she went and I closed the box, this time knowing I had captured the swarm. The knot of bees in front vanished at once, fading into the entrance to join their queen. "Fine," I said to myself as I put my glasses back into their case; but what about that bunch sitting on the other branches of the swarm tree? There were an enormous number of bees still circling in the air. Remember, I was thirty feet away. I carried the swarm box with me back to the swarm tree. I figured the bees would see the box, smell the queen and dive for the entrance as all the rest had done once queenie was inside. A few did, but I could see that communications between them had been broken when I moved them away originally.

I put my glasses back on so that I could examine

the small gathering of bees on one branch near the stump. Perhaps another queen was there. I knew this does occur sometimes. As I bent over the branch under the one I was looking at, I became aware of the number of bees that were flying around my head. I had removed my bee veil just after placing the branch in the box. After looking at the remains of the swarm on the branch, I decided that another queen was not present; so how to get the rest of the bees into the box? I reached up to my glasses again, this time to replace them into my case, which rested in the left-hand pocket of my sport shirt. As I looked down, ready to replace the specs in the pocket, I was shocked by the sight of thousands of bees busily diving into my glasses case. Even as I watched, they bubbled up, unable to fit into it and instead cascaded into a long flowing beard of bees, which in moments was long enough to reach as low as my knees. I'll take it any way it comes, I thought to myself as I gently lifted the glasses case, bees and beard out of my pocket. At the entrance to the swarm box, I crossed my fingers and shook! The bees tumbled to the ground in front of the entrance, turned toward it, and began to fan violently. Suddenly, those bees still in the air began to light near the fanners while others began to make their way into the interior to join the queen.

I don't believe that anything in beekeeping is more amusing than watching bees transmit their "come hither" message. Their little behinds are pointed skyward and feet are planted firmly while the wings rev up to an incredible number of oscillations per second; so fast that their wings are only a blur. The scent glands in their rear ends secrete a pheromone which says, "The queen is here, the queen is here!" Their wings send the message on its way. Of course, that strong pheromone, called queen matter, was on my hands when I replaced my glasses in their case. The scent stayed in the glasses case, and when I approached the bees in the tree this last time, they were sure their queen was in that pocket of mine.

About a half hour later, not a single bee was anywhere but inside the swarm box. Now I opened the top ventilation cover so they would have the advantage of air through the screening. At the same time I slid the screened slide into the entrance, thus containing the bees inside. A few fat drones who couldn't get in flew to the screen on top and traveled home with us as hobos. They wouldn't leave that nice meal ticket. The cover to the ventilation screen must be kept in place while the bees are going through the entrance; otherwise, the bees on the outside will head for it and try to get together with their sisters through the top.

When I reached the bee yard, I headed for a prepared hive waiting for this swarm. The hive body had ten frames with foundation, bottom board, and covers. I spread a white sheet on the ground in front of the hive, leading right into the entrance. If a sheet is used this way, it provides a rampway into the hive and the bees will not congregate under the hive. The cover of the swarm box is removed and some bees are shaken out of it, in front of the entrance to the hive. I usually leave the rest to the bees. The box is tipped over on its side so the open top faces the hive entrance. All that remains is for the beekeeper to enjoy the show, for the bees seem always to know

*Moving the captured bees from swarm box to new hive*

what to do. I'm sure they say to each other something like, "Smell that fresh wax in there? Let's have at it!" And do they ever! Loaded with honey, as they are, they can make wax so rapidly that half a dozen frames can be drawn into comb within a week. Always hive a swarm on sheets of foundation if you can. It helps avoid the possibility of importing a disease, since the honey, which may be carrying spores of AFB, will be used by the bees to make wax—a process which apparently destroys the disease—long before any brood are in the hive.

Upon returning to the bee yard, it occasionally turns out not to be a propitious moment to hive the newly acquired swarm. The design of the swarm box provides the answer, since the entrance slide may be removed and the bees allowed to go and come as they please. The only provision you must make certain of is to locate the swarm box where the bees will be permanently installed in a regular hive as soon as time permits. Naturally, the sooner they are hived, the more they can effectively use their honey cargo.

Now, there was the swarm we caught at . . . Hold it! You know, every beekeeper has his share of swarm stories. He'll spin out his favorite yarns and listen to some of yours, if you have a few to tell. When swarm fever starts your blood pounding this next spring, go get your share of stories to tell!

# 11

## *Some Common Mistakes and Some Helpful Tips*

In addition to learning how to cope with swarming, the beginning beekeeper should be aware of some common mistakes that, unless avoided, could reduce his success. Every beekeeper cries softly to himself from time to time. At least I think he does; perhaps I'm wrong, but the probability weighs heavily on my side. Beekeepers who continue to pursue their hobby grow with their experience; and experience means making mistakes and learning by them. Some errors, which can be costly in the beginning, can also make you cry. When your faithful workers die, sometimes *en masse*, because of your blunder, you'll vow never to let it happen again, and it won't! Some mistakes we make are common and some are unique. Let's try to prevent some of the more common ones from occurring, because a simple error could cause a failure for the season.

An entrance reducer is a valuable tool when properly used—far more important than most beekeepers realize. You will recall that when you hived the bees

*An entrance reducer*

the larger of the two openings on the reducer was used, together with the feeder. That left a "one-finger"-sized opening. It was early spring. Then, as the colony grew and the weather warmed a bit, the reducer was removed so the feeder could be retained; but you used grass, or made a cleat to stretch from the feeder to within four fingers' width of the opposite side. The entrance was still restricted because the colony still had not grown strong enough to defend their hive from invaders.

*The harmful effects of robbing bees on a weak colony*

Robbing bees create a serious situation. Not alone can they destroy a small growing colony, but they create an atmosphere of irritability that makes dispositions that were serene become nasty. That can mean increased numbers of stings. Syrup feeding needs great care in the application. A few drops of syrup anywhere but in the jar can touch off robbing because the scent can attract bees from other colonies. Great activity at the entrance to the hive need not mean that the colony is booming. Quite the contrary, it could indicate that an invasion is on and your colony is being stripped of every drop of accumulated food. Early in the spring is the period when robber bees are most active. Early in the fall, the yellow jackets (wasps) join bees in robbing weak hives. During these periods, nectar is not readily available, so their target can be any small colony. Unfortunately, this could be the death of the colony by starvation.

*Recognizing a robbing situation*

Learning to recognize a robbing situation brings us back to that business of experience. Watch a regular hive for fifteen minutes or so and you'll be able to pin down what a normal bee's takeoff from the front porch looks like. Some foraging bees walk out to the end of the bottom board and suddenly they are airborne. Others shoot out of the dark recess and are in the air before reaching the end of the runway. Foraging bees leaving the hive rarely climb up the face of the hive before departing for their nectar-gathering flight; but robbing bees do just that! Filled with honey, they are heavy and need that extra height for their takeoff. A downward dip in their trajectory is characteristic as they leave the front of the hive.

Foraging bees returning from gathering nectar or pollen land heavily, weighted as they are with their loads. In fact, many miss the entrance, and land on the grass in front of the hive.* Not so with the thieves. They are empty when they approach the hive and fly warily in front of the entrance, from side to side, awaiting their opportunity to slip in unnoticed. Careful observation can reveal these differences in flight behavior and result in early diagnosis of the problem. Reducing the entrance to a very small opening, usually the size of one bee, is the most expedient move to make. A few loose strands of grass over the opening also helps to dissuade the robbers.

*Preventing robbing*

If the whole matter has gotten out of hand and there is an uproar resulting in bees rolling around on the bottom board in mortal combat, then stronger moves must be made immediately.

Use masking tape to close off all possible cracks; even if a bee cannot get through the opening, the

*A friend of mine takes a three-inch strip of screening and tacks it on the bottom board across the full width of the hive. It hangs down by a couple of inches in front and helps the loaded bees that fall a wee bit short of the board to find the landing platform.

scent of honey or syrup can. Then cover the hive like a tent with a large sheet, after stuffing it in a pail of water to soak it thoroughly. The sheet will hang to the ground but any of the bees that live there will find their way in; the robbers hate to go under a wet sheet. Do not leave the sheet on for more than two days at a time.

*Using the entrance reducer*

The entrance reducer must be used also as a temperature-control medium. Judgment of night temperatures determines the size opening to use. All through the summer, the full width of the bottom board is used as an entrance. No reducer is employed in the late spring, once the night temperatures do not drop below sixty degrees Fahrenheit. When the first chill night at summer's end drops the mercury to fifty degrees, bring out that simple but very important block of wood, the entrance reducer.

There is no need to be sad if you don't make the error; but someone I know did. They left the entrance reducer on the hive all year long, using the smallest opening. That colony didn't have a chance!

The inner cover belongs with the outer cover. They travel together like a pair of gloves. The outer cover has but one use, to protect the hive interior from the weather, be it hot, cold, rainy, or snowy. Not so the lowly inner cover, which is multi-faceted in its uses.

*Using the inner cover*

There are three or four different models around, but I use an inner cover that has one flat side and a raised ledge around the perimeter of the opposite side. Grooved out of the ledge, in the middle of one end (the short dimension), is a half-moon type of opening. In the dead center of the flat portion of the cover there is an oval opening. This opening is the proper size to receive the metal bee escape used in the fall to clear the bees out of the surplus honey super that the beekeeper harvests.

The ledge side of the cover faces upward most of the time. It is on this ledge that the outer cover rests, forming an air space large enough to create insulation both winter and summer. In the winter, the inner cover also functions as an emergency supply of rations for the bees, since sugar is spread all across the top surface.

In the early spring (end of February on) a good feeding of pollen patties will develop the strength of a colony like a spring storm. It is then that the inner cover is reversed so the ledge is down, giving room for the patty on the top bars of the frames under it.

But with almost every other use, the inner cover sits flat side down on the four corner wedges I mentioned previously. Both openings are left open. When the outer cover is pushed forward from the rear, a fine upper entrance is created by the half-moon opening. Pull the outer cover back and the ledge opening is closed, should that be desired.

Together, those two openings provide for the colony ventilation that is constantly required. The bee's breathing apparatus consists of ten openings on each side of her thorax and abdomen, known as breathing pores. The inhaled air is passed to a series of air sacs which correspond to lungs. The air sacs are found throughout the bee's body, even in her head and legs. The air sacs turn into tubes that carry the oxygen to the cells via oxygen-saturated liquid. This entire mechanism functions because of a natural expansion and contraction of the air sacs brought about, in part, by muscle activity. The reason for

*Using the inner cover to ventilate the bees' efficient heating system*

giving this oversimplified description of the bee's respiratory system is to point out that the metabolism produces considerable heat. This heat is not held in the bee's body, but released by radiation. The hot air rising from the bees contains vapor, and as the hive interior temperature increases, so does the water-vapor content. The hive must have the means for upper ventilation in order for the interior to stay dry.

*Using the inner cover to separate two colonies*

The inner cover can also be used as a separator, either temporarily or semi-permanently, between two colonies, one set above the other. If you will screen both sides of the oval hole and cut an entrance an inch wide through the ledge on the side, it can function as a bottom board for the colony above. Early in the spring, you can form a small nucleus colony with a couple of frames of brood plus one frame each of honey and pollen. Place these in a deep body, which will use the inner cover as a bottom board and place it on top of a functioning hive; the new colony will use the heat generated by the colony below that comes up through the screened oval hole. It helps to incubate the brood above. See that the "nuc" has eggs or larvae from which the bees can make a queen, or add a purchased queen. In a month to six weeks you will have a nice young colony of bees. This "nuc" is different from one started in a "nuc" box, which we'll discuss later.

A summer or two ago, a beekeeper friend wanted to know why he had not come away with some surplus honey even after his second year? When questioning him, I discovered that the inner cover never was raised higher than the top of the first brood chamber. The second deep, or food, chamber was placed on top of the inner cover all season long. The bees swarmed repeatedly until the population balanced with the space provided. Certainly a mistake that cost him a honey surplus for two years. He was lucky that he didn't lose his colony.

*Misuse of the outer cover*

Other mistakes that appear to be common among neophyte beekeepers includes the misuse of the outer cover. Three different people came to me, in as many years, and told me that they only used the outer cover when it rained. "It's a rain cover, isn't it?" was the question put to me. I'm really not criticizing these folks, for it just happens they all were from California. Chances are it was just a natural thing for them to think; but please keep the rain hat on all the time.

*Other mistakes*

Additional errors and oversights crop up from time to time, but the most common blunders include frying the bees when smoking them; drawing the inner frames out first—thereby rolling the bees against each other and killing enough to make the colony furious; opening the hive as often as every two days, "just to see what's going on"; and a host of other unfortunate moves that cause trouble.

*Always think before you act*

One day, around the third year I had been keeping bees, I stopped just before carrying out a certain manipulation and asked myself a truly valid question: "In all respects, what exactly will my actions and moves mean to the bees? Precisely how will it affect them?" I have never carried out an operation with the bees since then without first making that query of myself. It has resulted in a large number of negative moves, printed boldly across the front of my mind.

*Never
leave syrup
or honey
around*

*Always
wear your
bee veil*

*Don't be
too nosy*

*Make sure
there is
adequate
space be-
tween out-
side frames
and side
walls*

I've learned never to leave any syrup or honey around, not even a tiny drop, when feeding the bees.

One time I was lazy and approached a hive without a head veil. I was zapped on the face. I decided I would never be lazy and would always wear the veil whenever I intended to do any work with any hive. Oh, I break the rule now and then, but I usually pay for it.

Once in a while, circumstances have urged me to stretch a point or two, when checking for the quantity of brood early in the year. Invariably I learned that those colonies I upset by being too nosy, too early, were set back sufficiently to run behind the others that I didn't bother till later. I don't mean for us to trust to luck and not check out the bees, but the rules are to be followed; if it is cool and there is no sun, leave them alone, in early spring especially.

A small matter, which became a large one with me, concerns the space left between the outer frames and the side walls of the hive. It is hard to realize that, unless a bee space is available at that point, one can curtail production overall by at least 10 percent. The outer sides of those outer frames are usually barren in most people's colonies I have visited; that is because they are too close to the side walls. It has become ritual for me to make a final move before closing the hive. The hive tool (curved part) goes between an outer frame and the hive wall on one side and my other hand crosses the hive to the outer frame on the other side opposite the tool. I squeeze the frames together, leaving proper space alongside each outside frame. Then I move the tool to the other end of the frame and my hand again is opposite to squeeze the other end of the ten frames. One caution:

move your hand across the hive slowly but steadily and the bees will ignore it; move it rapidly and it can attract them to zap it.

When the visit with your colony is ended for the day, and the cover goes on (rain or shine) always weight it with a rock or a brick or two. Though it is a rare occasion, the wind can and does lift covers off and usually during a rainstorm. My rock weights serve a dual purpose; they also indicate certain conditions within the hive. When everything is okay, they are placed in the center of the cover; when the colony needs attention, rear left corner. If I suspect the colony is queenless and needs more careful inspection, I add another rock on top—if time does not permit a further look just then. I also use the rock or brick to signal the amount of syrup left in a jar when feeding inside; far right corner means full, middle front is half full, and left corner signifies the jar is empty.

*Place a weight on top of outer cover*

I couple up these signals with a log book which has notes as to what I did last to a particular colony, with the date of the visit. Once the cover goes on a hive, it's amazing how difficult it is to remember exactly what you did and what needs to be done on the next visit. A bee master friend goes even further. He has a white cardboard tacked up inside the outer cover. He has ruled it into five or six columns which have the headings he requires, such as: queen, frames of brood, honey, pollen, etc. He also keeps a log, but when he raises that cover, he is able to make an immediate comparison to check progress of the colony.

*Keep a log*

Everyone develops a pattern of conduct that expands into a series of habits. Each action, therefore,

*Gloves and other clothing can attract stings*

performed by the beekeeper is pretty much the same each time. Even the improper moves are repeated, and unless the beekeeper has something or someone to compare his procedures to, he is very likely to repeat a less than proper mode of action. A good example concerns the use of gloves for protection from stings. Many hobbyists never graduate to working the bees with bare hands; but those who make it that far discover the main reason for their awkwardness was the gloves. The clumsiness seems to disappear when the gloves disappear. When wearing gloves, it appears each visit to the hive brings out nastier bees than the last time. The stings grow in number, making it difficult to work, and beekeeping becomes definitely less than pleasurable.

It is known that pheromones, or scents, are released at the site of stings. They broadly sound the alarm to the rest of the colony. The pheromone is powerful and the spot holds the scent for a long time. Any clothing worn by the beekeeper that has been stung without his knowing turns him into an attractor for more of the same. The gloves, in particular, fall into that category. They receive more stings than other areas since the beekeeper does not have a sense of "feel" when gloves are worn, and he squeezes bees which then sting. The next visit to the hive wearing the same gloves or clothes continues the difficulty, but on a higher starting level. It finally can become intolerable, though the solution is so simple. After working with the bees, smoke your gloves and any other clothing that you might habitually wear without washing prior to the next visit.

I recall an invitation I once received from a beekeeper who owned his first large bee yard. He wanted to be a commercial beekeeper. Dick asked me to visit his yard with him. He claimed that his bees were ferocious and that he and his helper had trouble even getting close to some colonies. After we had talked awhile about his yard, I must admit I was apprehensive and did not look forward to my role of expert. I did accept, however, and found myself at his bee yard a week later. Dick and his helper, John, met me on a country lane that led from a meadow to the bee yard.

I had decided not to be a hero, so I fired up my smoker and donned a full white jumpsuit that connected to my head veil with a zipper. After stuffing my pants into muscle-high boots, I was ready for anything. Yes, I had gloves too, but in my pocket. I just couldn't bring myself to put them on. I think I half felt I would be sorry that I didn't. I asked the boys which hives gave them the most trouble. They both agreed on two particular colonies. As I started up the lane for the couple of hundred yards to the hives, Dick put out his cigarette, as though he were going along. I explained that I would like to go it alone at first, but that I would call them to come in later. Dick sat down on a rock and lit another cigarette.

*Visiting an "angry" bee yard*

The colony with a neat 22 stenciled on the side of the cover looked busy, the bees leaving as though shot out of a cannon and falling all over themselves as they returned heavily laden. I watched for another five minutes, getting my nerve up. "After all," I said to myself, "they claimed they couldn't get within eight feet of this one and here I am close enough to use my reading glasses." The bees didn't seem to know I existed; but of course I knew that they knew I was there.

I smoked the hive in textbook fashion and re-

*No stings*
*for me*

moved the outer cover. The hive was made up of two deep supers. It was June and it was boiling over with bees. The inner cover was removed and a few bees walked my hands in the usual inquisitive manner, but the familiar hot flash didn't come. I threw a little more smoke over the top of the frames and went for number one. The end frame came out and so did a load of bees with it. This was a healthy colony with a full complement of boarders.

By the time I reached the fifth frame, the sweat ran down my arms and dribbled off my fingers. My glasses were steaming and slipping off my nose. I just couldn't take it, so I peeled off the straitjacket that encased me. I was a great deal cooler in suntans and a summer sport shirt. My pants were left in my boots and I wore a head veil.

I returned to the hive expecting some trouble. I was willing to take a few stings if the colony proved to be mean, but not swelter in the early summer heat. I gave them a little more smoke, saw the queen, counted the frames of brood, and decided they were healthy. As I closed the hive, I thought this might be a fluke; why not try another mean colony before calling the fellows? I moved over to No. 36, approaching it warily, for I still expected the worst. The second colony reacted much like the first. I began to think this might have been Dick's way of getting me over there to just check things out. I gave them a shout from the head of the tiny trail and they came on the double. I guess they thought I was in trouble. They both scrunched to a halt, looking like two moonmen in their full suits of white. I didn't know they were wearing full winter underwear under the suits to boot.

"What happened? Where are you stung? Are you

all right? Where is your suit, why aren't you wearing it?" They were firing questions the way a machine gun spits bullets. I gave them a rundown of the last twenty minutes, and even through their veils I could see puzzlement and consternation on their faces. I didn't know the answer either, but I had a hunch. "Let's take a look at another colony together," I offered, as I restuffed one pant leg into my boot.

We headed over to No. 23, Dick in the lead and John bringing up the rear. As we neared the hive, Dick slapped at the rear of his shoulder and said, "Ouch." I heard another "ouch" behind me and more slapping by the leader. I stood in the middle, watching these two men being stung through their bee suits and was positively bewildered. "See what I mean," Dick sang out between slaps. "This happens every time with nearly all the hives."

*Many stings for my friends*

We backed away and went down the path to the cars. It was only then that they realized I had not been approached by the bees, much less attacked the way they had been. They peeled out of their suits and then looked me over very carefully. Was I a stoic and wouldn't let on when stung? That seemed to be their suspicion.

I began slowly, knowing this would not sit well with them at first. "Gents, I have a proposition for you, if you're game." There was no answer, so I continued. "Put on your head veils and let's go back to the bees and try again." They both looked at me like I was crazy.

Dick snapped a stick in two between his big fingers. "Ed, you know I'll try anything, but it just don't make sense to go there naked and let them kill us," he grumbled. Dick is a big man, and he looked so funny in the hot sun wearing long white underwear. "What do you have on your mind to try?" tested John.

I asked them to get out of their underwear and put on their regular summer clothes that were in the car. First, though, I put some tobacco in my smoker and gently smoked the areas on their bodies where they had been stung.

*Removing their white jump suits*

They agreed and we all went back to the hives wearing head veils, trousers stuffed into the boots, and no gloves. I led the way and reached the "tough" colony No. 22. We stood shoulder to shoulder at the side and rear of the hive. The bees continued to work peacefully. I asked John to show me how he opened a hive. He smoked the entrance, yanked off the cover, meanwhile smoking the bees on the inner cover, which took to the air. The outer cover clattered to the ground, and the hive tool in his left hand slid under the inner cover, which snapped upward as his left hand on the tool went downward. Meanwhile, the smoker was pouring clouds of smoke (and a little fire) all over the hive. More bees took to the air. John dug out a frame without a vestige of care or gentleness. He was stung on his arm.

We closed up the colony and moved over to another hive. They were both smiling broadly, still not believing that it was possible that we weren't each stung fifty times.

*Virtually no stings*

At another colony, I went through the hive-opening procedure, explaining the steps as I went along. John tried the technique on the next hive and Dick worked open the next. All went smoothly except for one sting that Dick received on his right-hand index finger.

Back at the cars, I theorized on what I believed was

the problem. The suits had been stung a number of times and not washed or smoked; this just brought out the very worst the bees have in them. After all, the bees were just following their God-given talents to respond to an alarm. The gloves they wore were also responsible for alerting the bees. The long underwear caused more trouble as their overheated bodies began to get pretty rank from perspiration. John's entry into the hive was a proper procedure all right—to get bees as angry as they can be. I explained to Dick that it wouldn't hurt to remove that nicotine stain from his fingers and to refrain from smoking just before working the bees. Smoke is one thing but that acid smell on his fingers was something else. About three years have passed since then and Dick has about 150 colonies of bees. Handling the bees and bee stings are no longer his problems. *Accounting for the difference*

Of course, the dispositions of various strains of bees do vary. I believe it's better to have good-producing bees that are gentle, rather than good-producing bees that are too hot to handle comfortably. Hybrid bees are available that can fulfill these requirements, which are especially welcome to the hobbyist who is not experienced at handling aggressive bees. If you do go the hybrid route, be prepared to invest in a new queen every two years in order to maintain the same strain. Your virgin queens born in the hive will mate with any old drone who comes along, regardless of his ancestry. A crossed hybrid is the result, and the bees produced are, most often, cross in disposition; but it generally takes two or three generations of crossing to make this occur. The advisability of talking with a friend who has been keeping bees is obvious. If he has had hybrid bees, *The unreliable dispositions of hybrid bees* the advice will be valid; but if not, keep looking for someone who has had the experience. Then you can weigh the advantages against the disadvantages.

*Caucasian bees are mild*

Caucasian bees, for example, are notorious for making propolis or "bee glue." Well, Caucasians were used in the making of one of the hybrids and, yes, the hybrid does propolize a bit more than other races of bees; but the Caucasian is a very gentle bee and so is the hybrid that comes out of it.

*A new hybrid from artificial insemination*

Recently, a new hybrid was offered to beekeepers. This hybrid, as are most, was developed through the use of artificial insemination. In this way, complete control is accorded the bee breeder. Desirable characteristics are carefully selected from various races of bees and combined into one hybrid. Drone control is afforded the geneticist only when artificial insemination is employed. Since mating normally takes place in the air, there would be no way of knowing what kind of drone impregnated the virgin queen. The new hybrid is still only referred to by a number Cale 876. I must admit to trying a couple of these special queens. They appear to have outstanding qualities, and the colonies I gave them to seemed to agree. I have never seen such a rapid buildup. It remains to be seen what happens during next summer's production of honey as well as how they will overwinter. There are many factors which must come together affirmatively to make for a wholesome and productive new hybrid. A look into future bee-breeding programs is both promising and exciting.

# 12

## Summer

Finally there comes a stretch of days when the air turns warm and the blue sky holds the sun longer. It is summer and the golden dandelion says so as the blossoms disappear. Now the hive pace quickens, sometimes to a frenzied tempo, as each worker bolts from the doorstep of the hive, airborne like an arrow to the blooming flowers for their gifts of nectar and pollen.

The colony's population has exploded and now is five or six times its original size—if it has not halved from swarming. Now they plunge all out to gather stores of nectar all day, nectar which will be stored to ripen into honey. The bees labor now, outside every day and inside every night, for they know in the short space of time ahead nectar flows in abundance, and that suddenly, even while summer heat makes its furnace, the main flow will end. Then they must wait for lesser, more moderate sources of nectar that will ebb and flow till summer ends and autumn comes with her goldenrod and asters.

*The period of the crucial honey flows*

An acquaintance of mine thought that the term "timing" was underrated. He felt that everything

*The importance of timing*

required timing and no one ever really thought too much about it. He said that the people who applied "timing" to every equation of life were always contented, successful in their endeavors, and most of all, smiled a great deal. I guess he knew what he was talking about, for he ran an airline. He probably would make an excellent beekeeper, because timing must be carefully considered in all phases of the apicultural art.

*Adding shallow super*

As a new beekeeping enthusiast, you have been in attendance to your colony of bees since hiving them in mid or late April. Now it is June, and with the care you have given them, the second-story food chamber has been filling rapidly with brood and food. You haven't stopped feeding syrup until now. Suddenly there seems to be too little room for the multitudes, and you look carefully into the hive and find it is true. Only two or three frames are left up top, and it is time now for them to provide you with a surplus. A shallow super is needed to hold the honey that will probably come into the hive between now and the fall.

*How to tell if you need a queen excluder*

The next question that will come to mind is "Do I need a queen excluder or not?" The answer: Well, yes and no! It depends on the situation and the timing. Situation: The queen began egg laying in the brood chamber. She filled all the cells with eggs as fast as her workers created them from the flat sheet of wax that they were given. You observed her progress, and when the majority of the frames (seven out of ten) were filled with something, be it brood, honey, or pollen, you doubled the size of the building by adding story No. 2, the food chamber. The queen continued laying in the brood chamber as long as there was room, but the nectar and pollen crowded her out. The bees had drawn some comb, using the syrup you continued to feed, and there was room for her above. So up she went and plugged the cells with eggs. Meanwhile, the brood in the lower chamber hatched out into adult bees during the three or four weeks after the queen went above to continue her egg laying. The cells they left were cleaned and polished, and even though syrup and nectar were being stored and processed, the dark central areas of the frames, where brood had hatched, were not used. Some of the newborn bees joined the queen in the upper chamber and, functioning as nurses, speeded events along. The growth of a colony is like an explosion, once it is rolling ahead; but now it is your move. You are considering the shallow super to be added and the advisability of using a queen excluder. As a hobbyist beekeeper, you do have options which the commercial man does not. Nearly every honey producer automatically adds an excluder to every hive at a certain time of the season before adding the surplus chamber. I repeat, you have the option, he doesn't.

You'll know what to do after inspecting the upper deep chamber. Search for eggs, which are evidence of the queen's presence during the last two days, or find the queen herself. If you do find her or the eggs, you need the excluder. Should neither be present, it means that your efficient queen decided to use the space below, provided for her by the hatched-out bees. Look carefully, then, at enough frames in the upper deep hive body to assess how many empty cells of comb are available. At this time of the summer (June and early July), two or three empty frames can be filled in as many days. You are therefore safe in adding the shallow super without the excluder between.

Why all the fuss about its use? The excluder serves a very necessary function by keeping the queen out of the honey supers. The combs stay light in color, and the lack of brood in them also keeps the pollen storage below with the brood. Unfortunately, the excluder slows the bees down as they pass through it. It is a barrier of sorts, even though they can go through it easily. Many bee people think that a bee filled with nectar is too stout to pass through it. Actually, the bee swells but little with nectar; rather her segmented abdomen elongates, somewhat as does an accordion.

The time of the season requires you to watch the filling of the shallow super, for it can happen quickly when an intense flow is on. Certainly check it weekly. In your first year's use of a shallow super, the bees must make comb out of foundation; but if comb is placed above, the filling of the cells with honey is greatly hastened. In your second year of beekeeping, the comb will be available for you to use, after you have extracted the honey from it. The comb in the frame is reusuable. Normally, harvest time for beekeepers corresponds to harvest time for gardeners and farmers; but with my own bees I am finding it better to move the harvest period up.

The spread of the cities and suburbia has somewhat curtailed the activities of the remaining farms that were once not too far from the city. These farms now either no longer exist or have been reduced in size by local developers' building of homes. Previously, these farms provided a fine flow of nectar as one crop or other on perhaps five to fifty acres came into bloom. All you needed was enough bees to bring it back. Now the same areas depend upon local vegetation, usually in the form of those misplaced flowers, the weeds. Weeds are not to be ignored as a source of nectar. Their list is varied and long. Naturally they will be different from one location to another. We do well with several kinds, such as dandelions, clover, various grasses, jewelweed, loosestrife, ironweed, goldenrod, and asters, to name a few.

The flows are not strong, though constant; as a result, I suggest removing surplus shallows when they are full and capped. If the season turns sour, after the initial flow, you will have honey. "What about the bees?" you say. They will find enough nectar to keep themselves fed through the summer. If they do not gather enough to carry them through the winter, supplemental feeding will rectify the situation. I'll cover that in chapters 13 and 15. *Harvest your honey whenever surplus supers are full*

Each season of nectar flows is different. Nature's conditions influence the changes that occur. Heavy snows or early spring rains usually provide good early flows. The sun must play her part too, or else the blossoms will not come. If a dry spell persists for too long, nectar availability will diminish. If the rains come at the wrong time, the flow can be a washout. We are farming, and that which affects the farmer does affect us also. So take your honey from the hive, when the super's frames are filled and capped and the strong nectar flow is over. Of course, while the nectar is readily available to the bees and they continue to return from the fields with heavy loads, leave the honey there and build a skyscraper *The effects of weather on honey flow*

*A super is
ready when
80 percent
of the honey
is capped*

of supers, one at a time, as they fill.

Many new beekeepers have a problem knowing when a super is ready to take off the hive. Frequently they are puzzled when the cells are full and wet but not capped over with the pearly-white beeswax. The rule of thumb has been to take the super off if 80 percent of the honey has been capped. The uncapped material will blend with the capped, or ripened, honey. Sometimes I am faced by this predicament and I solve it by being sure the uncapped honey is really honey and not nectar. I remove a frame that has uncapped as well as capped honey and, standing over the open hive, shake the frame toward the top bars of the frames below. If it is honey, nothing will come out; if nectar, or honey not yet sufficiently cured, the bees get a sweet shower bath. Nothing is wasted since they go right to work and lick it up.

*Adding
additional
supers*

When the flows are good and the skyscraper builds, you can find yourself with three, four, and even five shallows on at one time. The science of supering has it that as each new super goes on it should occupy the bottom position under all the rest and just above the second deep hive body. Theoretically, the reason for this maneuver is to place it as close to the entrance as possible, to minimize travel for the bees. Theoretically, that is correct; but it is difficult for the average beekeeper to lift off everything else in order to put one more super underneath. "Top supering" is the answer. When needed, add the next super over the previous one. The bees will be slightly slower in developing it, but you won't have a strained back. If the gods of nectar are with you and you are ready to add your third super, then help the bees by drilling a hole in the face of the shallow super under the hand hold. Any size from three-quarters to one inch is okay. Just be certain to have dowel material or a cork sized to fit the hole when you want it closed. The bees will be quick to find this upper entrance, and it is direct to their storage area.

One quick word about your choice of the form of honey you will produce. There are basically three kinds: regular extracted liquid honey, chunk or comb honey cut out of the frames, and section box honey, which is comb honey that the bees put into little boxes furnished by the beekeeper. The last one is the most difficult to produce. It requires a very strong flow plus an expert beekeeper, since the bees detest having to work in the boxes, and frequently desert the hive by swarming. Comb honey, cut out of the frames and attractively packaged, is in demand by everyone who knows good honey. The main drawback is its cost, since each time you cut it out a new piece of wax foundation must be furnished. It also takes time for the bees to draw out the foundation into comb, besides using some nectar or honey to make the wax. Comb honey, however, is so delicious—and it is natural. That leaves us with extracted honey, the conventional method of harvesting honey.

*Decide if you want liquid honey, chunk honey, or section honey*

As midsummer approaches, the weather generally turns dry and hot. The bees can use the beekeeper's help now. Shade the hive with wood slats over bricks placed on the cover. Also provide them with water. I mentioned chicken waterers earlier. I believe they are easy and convenient for the bees to use. Be sure to place small rocks in the rim of the waterer, so that the bees do not drown.

*Shade and water your bees in hot weather*

If your particular location places you where the humidity reaches 70 to 80 percent for a period during the summer, then be not alarmed when your bees sit outside the hive on the porch, or even form a beard clear up the face of the hive, for all or most of the night. Heavy, humid weather exaggerates the humid condition within the hive; and the bees seek relief in the cool night air.

*Tips before you leave on a summer vacation*

Summer is conventionally the season for vacations. If the mountains or the beaches beckon to you at this time, you need not fear that your charges will suffer even a little by your absence. They are too busy dashing about the countryside seeking every floret's gift of nectar. Give them one or two extra supers before leaving for your holiday. If the flow is a strong one, they may surprise you with extra supers bulging with honey. If not, you only have to remove them when you return.

About a week before your vacation begins, there is one very important preparation sometimes required. You must determine that your hive is queenright. I'm sure you have been monitoring the colony for this condition all along, but when you are going away for a few weeks, you court disaster should the colony be queenless. In case this sad event has occurred, there will be time enough to procure a queen and introduce her to the colony, if you inspect the colony one week before your vacation. Do not wait until one or two days prior to departure.

*How to introduce a new queen*

Queen introduction is a subject that has been discussed among beekeepers for an eternity. There are as many opinions as to how as there are beekeepers. If we are replacing a queen who has already died or disappeared, the matter is somewhat simplified.

Part the frames and place the queen cage about a third of the way down, in between two frames, with the candy end of the cage facing upward. Be sure the screen of the tiny cage is available to the bees who will feed this new queen until they free her by eating a passageway through the queen candy. If some of the attendant bees die in the cage, they will fall toward its bottom, permitting the queen full access to the passageway above, after it has been cleared of the candy.

If a queen is being replaced because she is inferior, then she must be located and removed from the hive before the new queen is introduced. It is not necessary to allow any time to elapse between the removal of one queen and the introduction of another. This is contrary to the procedure advocated in the past. "Leave the colony alone for at least three or four days, until they know that they are hopelessly queenless" was the modus operandi followed by all beekeepers. That usually caused a problem to develop. Sometimes the bees had a few eggs or larvae to work with and started a queen cell working. The introduced queen had little or no chance of being accepted under those conditions. On the other hand, if they really are hopelessly queenless for too long, one or more workers will assume the role of queen and proceed to lay eggs. Unfortunately, these eggs can only become drones, since our ersatz queen, being an unfertilized worker, can only lay drone eggs. This condition portends the death of the colony, for they will raise only drones until all the workers have lived out their life span.

When a colony has laying workers, it becomes impossible to have them accept a new real queen.

The laying worker queens will kill the newly introduced queen. Therefore, the make-believe queens must be removed before a new queen is released. Accomplishing this task is difficult, since these laying workers look just like any other workers. One method that seems always to meet with success requires the hive and all of its population to be transported at least a hundred yards away, preferably into some tall grass. A wheelbarrow is helpful at this point. Return the bottom board to the original stand. Now for the heart of the operation. Return to the bees and one by one shake all the bees off each frame. Every bee must be removed from the frames. If one bee remains, she might be a laying worker. Replace the frames, without any bees, into an empty hive body with a cover. Let no bees get into the super or on the frames. When all the bees have been removed, return the hive body and frames to the bottom board. Many bees will precede you to the bottom board. They are the inhabitants of the hive. The laying workers will be left behind in the tall grass and won't find their way back, since they were nurse bees and have never been out of the hive. Now you can introduce your new queen.

Anytime a new queen must be introduced, it is important to see that nary one queen cell is anywhere in the hive. It is necessary to remove the old queen from the hive. If she is in good health and is merely a low-volume layer, then she is a good queen to head up a "nuc," which every beekeeper should keep on hand through the season. A "nuc" or nucleus is a small colony of bees comprised of no more than five frames, usually four and sometimes as few as three frames. A nuc gives residence to a queen, held in reserve, should the regular queen die or be injured. This little nuc frequently grows into a full-blown colony by the end of the season, and the following year, will be one of your strongest colonies. The poor queen will sometimes be superseded by the bees in the nuc and a new fresh queen is the result.

Nuc box construction is rough and ready. Cabinetwork here is wasted. A kit is sometimes available, but if you aren't able to locate one, a simple box to hold four or five frames is easy to make. Add a bottom board to give clearance under the frames, and a cover to protect them from the weather.

A single frame of brood, which also has pollen and honey, with the attendant bees aboard, will be sufficient to care for your reserve queen. Of course, fill in the rest of the space with frames of foundation. If you want this nucleus to turn into a small colony by the end of the season, then start it with three frames of brood, honey, and pollen. Restrict the entrance to a size that permits entry of only one bee at a time, or else any strong colony will rob the nucleus of stores. When I establish a nuc, I close off the entrance entirely for at least a day and a night. In this way, the bees and queen establish the nuc as their home and later, when it is opened, they remain with it. Enough air is provided when a one-frame nuc is established, but if three or four frames of bees are used at the start, provisions must be made for a screened area to admit air.

A nuc is valuable, and is highly underrated by most beekeepers. As I mentioned, a reserve queen will have a place to stay, all the while functioning as the head of a colony. Too, this small colony makes

*Joining a nuc with an old colony*

queen introduction very simple. If requeening a colony is necessary, then all that needs to be done is to remove the old queen, if she is on hand, and join the nuc colony to the other colony over a sheet of newspaper, as you would join any two colonies of bees. Remove the inner and outer covers from the main colony. Spread a single sheet of newspaper over the top of the hive. Then make a few holes in the paper with a small nail or paper clip, before placing a hive body containing the frames and bees of the nuc on top of the hive to be requeened. The newspaper method of joining colonies is foolproof, since the bees from the two colonies have opportunity to join each other gradually, with an intermingling of the scents. The queen is almost always eagerly welcomed to the queenless colony, since she usually is bringing to the hive pollen and honey; and most of all, a complement of brood.

These maneuvers and manipulations are but a small part of the many interesting events that will take place during a beekeeper's first summer with a colony. Your mind will be bombarded with many great ideas as to how to do this or that. Jot them down on paper even if you cannot execute the idea at the moment. Later on, through the winter, include them in your plans for next year.

# 13

## Harvest Time
## and Fall Preparation

When the frothy white cappings on the frames of honey have been completed, and the signal for the removal of supers from the hive has reached you, do not delay. This is the treasure that you have sought. There will be competition from those who fashioned it, for they are reluctant to lose this booty they collected; so they must be cleared from the honey super before it can be removed. It is a simple matter to accomplish, though a choice of methods is available.

*Remove bees from surplus honey super before collecting*

The most common technique requires the use of a small piece of metal equipment called a "bee escape." When inserted in the oval hole on the flat section of the inner cover, which will be placed between the top super to be removed and the one below it, it provides a one-way passage from the upper super to the chamber below it. Once in position it usually requires twenty-four to forty-eight hours to clear one super of bees. There are precautions which must be observed.

*Use a "bee escape"*

If you do not have an extra inner cover to set up as a bee escape, then, of course, you have to remove the inner cover from the hive first. This is done in the usual way, employing smoke to quiet the bees. Replace the outer cover as soon as you have the inner cover in hand.

*Remove inner cover, insert "bee escape" in center opening and insert cover between top two supers*

Again, the inner cover is used flat side down and the escape is put on the side with the ledge running around as a frame. When you are ready to put the inner cover in place with the bee escape, smoke the bees in the usual way, especially between the super to be removed and the one below it. Use your hive tool to carefully pry the honey super completely free of the one it rests upon. Raise it slightly with the hive tool, while you pump some smoke through the crack. Do not rush; instead, wait a full minute before proceeding.

Now that the minute has elapsed, stand to the rear of the hive and, with your left hand in the rear hand hold, raise the rear of the super three or four inches. Your right hand guides the inner cover (ledge side up) in between the two supers. Lower the honey super onto the cover, meanwhile putting your chest or your knees, whichever is closest, against the end of the cover sticking out. Pull the super back, sliding it toward you and at the same time exerting some force against the end of the cover touching your body. It is a matter of pushing the cover and pulling the super. They will suddenly line up in the rear. Now the whole works (cover and super) is pushed forward until it lines up on the hive. Next, close the hole in the ledge of the inner cover with the half-moon piece of wood provided or with a plug made of grass. The outer cover is kept on the super all the time this is going on, so few, if any, bees are flying during this manipulation. Be absolutely certain that not a single bee space exists anywhere between the inner cover and the outer cover. The super must be tightly sandwiched in between the two covers, with only the bee escape as an exit. If even a tiny space allows a bee to re-enter the honey super, you will never clear it of bees, for they will go out one way and come right back through whatever opening exists.

A variation to the bee escape in the inner cover is called a ventilated escape board. It is the same as an inner cover in shape and size but screening is used inside the frame instead of wood. In opposite diagonal corners metal bee escapes are inserted to provide the bees with exits downward to the super below, and outward, through the frame rim, to the air outside. It works more efficiently than the solid cover, since the bees see their sisters below and seem to want to join them. It also is much easier on the bees in warm weather, since they are not shut up in an airtight box, as with the solid cover. They also see the light of day through the escape side exit and are attracted to it. This ventilated escape board is used the same way as the solid one. *(See p. 167.)*

*A ventilated escape board*

I am frequently asked why the bees have not cleared out of this or that super of honey after nearly a week of employing a bee escape. Usually it turns out that there was some brood in the super. The bees will not leave that brood, so you may as well wait for it to hatch before trying to get them out. If the bees have not cleared out of the honey super at the end of forty-eight hours, it will be necessary for you to check the metal escape to see if a particularly large

*Bees will not leave if there is brood in surplus honey supers*

drone might be stuck, blocking the way through it. The only other way the bees seem to "stay" in the super I mentioned a moment ago; a space was left for them to return to the super.

*Remove super and cover it*

After the proper time has elapsed, remove the super, but leave the escape in place. The bees will be contained in the hive by the bee escape until you are ready to proceed further. Place the super on a piece of plywood and, as fast as you take the cover off, replace it with several sheets of newspaper. If the honey super is not totally enclosed, and quickly, the bees will get into it from the air and all of your work will have to be repeated.

If nothing further has to be accomplished, puff a little smoke through the bee escape to the bees below it in the hive. After a few seconds, pry it loose, remove the metal escape and the half-moon block, and replace the outer cover.

*Repeat process if there is more honey*

If you still have more honey supers to remove, leave the items mentioned in place and repeat the insertion process below the next super to be cleared. Make certain to take honey supers indoors as quickly as they have been removed from the hive. If left out of doors for any length of time, a few bees will creep inside and cause complications for you during the uncapping and extracting procedure that follows.

*Other methods of removing bees are less kind to bees*

Removing the bees from honey supers with the bee escapes is but one of many different methods. Bee blowers have been devised that push the bees out with a blast of air directed between the frames of supers placed on their sides. Obnoxious chemicals that are foul smelling are placed on boards above the honey super, driving the bees out and down by their very odor. Frames of honey, lifted out of the super, are brushed clear of bees and then placed in another super on a table next to the hive. All of these systems work and will rid the super of bees; but I prefer the bee-escape method. It is so much easier on the bees and I think that is important. Somehow, I feel that it is more important now than probably at any other time. These tiny selfless creatures have labored constantly to provide us with a gift of nature. They are sharing with us a portion of their most highly prized possession; it surely is not too much for us to treat them with gentle kindness and consideration for their well being and an empathy for their point of view.

The honey supers are taken from the hive and carried to a closed area, as soon as the bees have been cleared from the combs. Once indoors, the urgency is over. It isn't necessary to uncap the frames and extract the honey immediately; but don't procrastinate too long, for if you do, the honey can granulate in the comb (how soon depends on the floral source of the nectar), and then it becomes impossible to use the honey unless the combs are melted down. This requires heat and then separation of the wax and honey. So get on with the job at your earliest convenience. Besides, it is one of the great pleasures of beekeeping, this business of watching the honey cascading from the spigot at the bottom of the extractor. I'm duty bound to call some little boys in our area, when I first extract the crop. They stick their index fingers into the stream of gold that the "candy machine" gives forth.

*Take honey supers inside*

The processing equipment required by the beekeeper will depend upon the size of his operation, that is, how many colonies of bees he has. Fortunate-

*Processing equipment includes heated knife, to remove cappings, and centrifugal extractor*

ly, this wondrous hobby permits the one-colony owner to adapt certain implements of the home for processing a single super of honey. I suggest these alternatives to minimize expense which is not warranted for harvesting one or two supers of honey. It might take a little longer and be a bit more difficult, but the end product will be the same. The first need to improvise will come right at the beginning. The wax cappings on the comb must be removed before the honey in the cells can be spun out by centrifugal force in an extractor. In order to accomplish this task, two items of equipment are used. The cappings are removed with an electrically heated knife. The hobbyist can get by the first year with the household's electric carving knife. If that isn't available, an ordinary bread knife heated in very hot water will suffice. The second item of equipment is a container to hold the cappings as they are removed from the comb. Some honey comes with the cappings and it must be separated from the wax. The container is referred to as a wax separator. There are many different types and models, some quite elaborate. A household colander placed in a larger container will do the job nicely. The wax cappings will fall into the colander, which allows the honey to drain into the vessel beneath it. A homemade separator can be fashioned from an old-fashioned round washtub with a wire basket placed in it to receive the cappings. Another system is to use a shallow super with a queen excluder nailed to the bottom. Support this unit over another container that will receive the draining honey.

I shall assume you have one of the suggested separators ready to be used and a knife has been made available. Hold one of the combs with the end of the top bar resting on the rim of the colander. I shall assume further that this is the item you settled upon. Incidentally, line the inside of the colander with cheesecloth that has been dipped in hot water and wrung out. The draining honey will pass through the cloth more easily if wet than if left dry. Lean the comb over the colander and, starting at the bottom end, use a back-and-forth sawing motion to cut away the cappings, which fall into the colander. Always cut the cappings from the bottom up so they separate from the comb, rather than fall onto the comb, as they would if one cut from the top. Use the top bar as a guide for the depth of the cut; but go deeper if required to remove the cappings from low spots. The bees will rebuild the comb, don't be concerned about minor damages to it. Treat the other side of the comb in the same way.

*How to remove cappings*

As each frame is uncapped, place it in one of the baskets of the extractor. "What extractor?" did you say. Well, it is an extractor which you rented from the local nature center or from your bee-supplies dealer. You could even consider borrowing one from a friend this first year. As soon as the extractor is filled with frames, you start extracting by turning the handle. Turn slowly so the cells will empty somewhat without the inside cells of honey pressing too strongly on the mid-rib of the comb. It is possible to rupture the combs if too great a strain is put upon them. After a few minutes, remove the frames and replace them in the extractor, but with the full cells facing the outside. Turn the crank again and the "rain on the roof" sound will provide music for your ears, as the honey pelts against the side of the

*Insert uncapped frames in extractor and spin it*

*Replace empty frames in super and return to hive for bees to clean*

extractor. You can develop a greater RPM now, since the inside cells on the frame were partially emptied and the pressures on the mid-rib are not as great. Still, use caution so the frames will not be broken. Continue in this manner until you have finished all the frames. Replace the empty ones in the super, since when you are completely finished the super with the wet combs will be returned to the hive, enabling the bees to clean out the last bits of honey. If the super is returned late in the afternoon, the possibility of robbing will be minimized.

*Strain honey through cheesecloth*

The honey in the extractor flows into a pail set beneath the spigot, but be certain to use two layers of cheesecloth over the pail to catch bits of wax, bee bodies and small pieces of wood carvings inadvertently whittled as you cut cappings. Remember to wet the cheesecloth and also be sure to secure it around the rim of the pail.

*Melt cappings into wax block*

Allow the colander to stand a few days while the cappings drain. That honey from the cappings, somehow, is always the most flavorful. When all, or most, of the honey has been drained, gather up the wax cappings and prepare them for melting into a block. Wash the wax, in the cheesecloth, to remove whatever honey remains. Meanwhile, have a pot of water brought to a boil, then turn down the heat so it just simmers. The pot should not be more than half full of water. While the water is simmering, not boiling, put the wax into the pot, a small amount at a time. When it has all turned to liquid, turn off the burner under the pot and let it cool down. Do not leave the stove until now, when the burner is out. If the phone should ring and you must answer it, put out the burner before you leave the stove. Beeswax is highly flammable and you can burn your house down ... if it boils over. Beeswax boils at a lower temperature than water. After the wax has cooled, you will find it has formed a block at the surface, dropping the dirt and impurities to the bottom of the pot. If you want to refine the wax further, merely repeat the process in a fresh pot of water.

After the bees have cleaned up the wet combs, they should be put away for safekeeping until next year. They must be protected from the wax moth larvae, which can destroy the combs if they get half a chance.

Fill the super with the frames and place it on a pad of newspaper. This will seal the bottom of the super from the air. Place on the top bars a sheet of paper about four inches square. A tablespoon of P.D.B. (paradichlorobenzene) is spread on the paper. Cover the top of the super with several thicknesses of newspaper and top it with a piece of plywood. If you have an extra hive cover, use it in place of the plywood. Several supers are handled the same way, each with its supply of P.D.B. Stack as high as it is convenient to reach. Top the entire stack with a cover of plywood. In spring you have only to air these supers for a couple of days before using them. Do not use naphtha flakes or naphthalen or any kind of regular mothballs. Most of these chemicals are absorbed by the comb and never lose their smell. The bees will not use them and then the wax must be replaced. Not so with P.D.B., which loses its odor on the comb quickly.

*Store cleaned empty supers for winter; cover with newspaper, sprinkle on P.D.B. to prevent wax moth, and cover with plywood*

Little is left to be done now except bottle the honey in any package that suits your fancy. Old-fashioned bottles, mason jars, and regular queenline,

*Bottle your honey*

or standard, honey jars are but some of the containers that can set off your honey handsomely.

Even as I accomplish the joyous task of harvesting the last of the surplus honey, summer signs decline and vague urgencies filter through my head like lazy smoke curling from a darkened fire. I am overvulnerable to my plaguing conscience, which reminds me not to commit the dalliance of previous years. Somehow, though my intentions make my mental lists complete and winter preparation is assured, I find myself always one short step behind completion of the task.

*Preparations* *for winter*  Each golden October I attack the list a little earlier, till now, I find myself drawing lines through the series of steps required for wintering as early as late September. There are many things to do; and they are multiplied by exactly the number of colonies. I sometimes say to myself, "There are roughly ten steps or items to accomplish and that's about a thousand actions for all my colonies."

The back-lot beekeeper can have things under control, since he does not have that many hives to give attention to. But start early and be finished with your obligations to your hard-working bees before it becomes urgent. There are certain things to do which I'll set out here in the order one would generally accomplish them; then I'll take the topics into greater detail.

1. Remove the surplus honey
2. Estimate the amount of honey in pounds you will leave the bees for winter
3. Inspect hive and colony for queen, brood, and disease
4. Powder medicate

5. Make up medicated sugar syrup
6. Inspect inner cover for ventilation wedges
7. Start inside feeding on inner cover
8. Sprinkle sugar on inner cover
9. Sweep floor of hive for mice
10. Substitute metal mouse guards for wooden entrance cleats
11. Drill top entrance hole if required
12. Limit bottom entrance to suit preference
13. Use moisture-releasing board (Insulite) on inner cover
14. Single-thickness tarpaper wrap to absorb heat of sun
15. Determine adequate windbreak

Helping your bees live through the winter is probably number one in importance, even before preventing swarming. After all, if you lose your entire colony, that is twice as bad as swarming, where you only lose half. An average-strength colony can make it through the winter if adequate stores of food are available to them in the hive. They also require adequate ventilation and sufficient shelter from prevailing winter winds.

*Sufficient food, ventilation, and shelter are crucial*

A mild-weather stretch in late fall may fool you into thinking that winter might not come. It surely will, and if you did not prepare your colony, you might lose it over the cold months ahead. If your colony is in a mild area toward the south, then you need only see that adequate stores of honey are there for the bees.

Most beekeepers do not realize that the bees rarely die as a direct result of cold weather; but instead, they starve to death because of long periods of frigid

*Bees die from starvation, not cold*

temperatures. In the two deep hive bodies, the bees come together and form a ball or cluster, mainly in the lower chamber but extending upward into the bottom portion of the upper chamber. This position accommodates the bees' need for communication since the cluster embraces the bottom of the upper frames and the top of the lower frames. The organization of the bees insures their contact with honey as the winter months pass, since the cluster moves slowly upward, eating their prepared stores of honey as they go. It is their inability to move on as a body to new stores of honey both above them and laterally that brings about starvation with honey perhaps an inch away. If the mercury rises to even approach the freezing point, 32 degrees Fahrenheit, the bees have a degree of mobility.

*Bees cluster for warmth and maintain 93° at center*

The queen is contained in the center of the cluster, where she will commence egg laying in middle January. The center of the cluster is maintained at 93 degrees for her well-being, as well as to provide incubation for the new brood in January. The queen's production is geared to the attrition through the winter. She cannot make up for excessive losses, however, that may be due to old bees going into the winter.

*Cluster's heat rises to inner cover and produces moisture*

The cluster's heat rises until it reaches the inner cover, which is considerably cooler. This causes moisture to form on the under side of the cover and eventually drip upon the cluster as ice water. The outside of the cluster is a three-inch band of bees that form an insulating shell maintaining approximately 45 degrees even though air temperature in the hive may drop to zero! As long as they stay dry, they can function; but when the moisture gives them a cold shower, their chances of making it through the winter are seriously jeopardized. This can be entirely avoided by affording the colony proper ventilation. The inner cover must be positioned with the ledge side facing upward and the flat side down. The oval hole in the center is to be left half open all through the winter. If your inner cover has a ventilation hole in the rim, leave it open. If yours does not have a hole, cut a half inch out of the rim. Save the piece, in case you ever want to close the hole thus created. When the outer cover is placed on the hive, push the cover forward (when standing behind the hive) so that you will be certain that the ventilation hole in the rim is open. If the cover is pulled backward, it will close the hole.

As described on page 15, in order to provide a narrow air space between the inner cover and the hive body, a spacer, about the thickness of the thin end of a shingle, is used. Place postage stamp sized squares of wood at the four corners on the flat side of the cover. The crack for air should not exceed 1/16 inch between the body and the cover. If larger, the bees will take issue and seal it with propolis. They will leave the majority of the small air space alone.

*How to ventilate*

When the outer cover is placed on top, in addition to pushing it forward, be sure the space on the sides is equidistant, in order to permit air currents to pass under the inner cover from one side to the other. Use your fingers to gauge the distances, for if one side is closed off, you will not have an airflow which carries off the moisture found on the underside of the inner cover.

The front entrance must also be left open, though considerably reduced. A metal mouse guard can

*Check for mice and install mouse guard*

eliminate another winter problem if the hive's bottom board is swept with a curved wire coat hanger, before the guard is fastened into place. The coat hanger will "feel" an obstruction if Mr. and Mrs. Mouse have set up housekeeping. Recent studies reveal that a totally closed bottom entrance slows air movement, holds down condensation and reduces food consumption. In this case, a top entrance auger hole must be provided.

*Fall feeding with sugar syrup 2 parts sugar to 1 part water*

It is time to bring out that gallon jar again to feed a mixture of sugar and water to supplement the bees' winter stores. This time, however, a mixture of two parts sugar to one part water will feed the bees a thicker syrup that will give them less work prior to storing it. The first two gallons fed to them should be medicated as in the spring. The difference between spring and fall feedings follows functional patterns. In the spring, feeding serves primarily stimulative purposes. The queen immediately becomes more active since the syrup is like a light nectar flow at that time. The bees build comb, and the queen lays eggs. If stores are low, this stimulative feeding can prevent starvation until nectar and pollen are available. The feeding must be kept up once started.

*Fall feeding can prevent winter starvation*

Autumn feeding does not provide a stimulus, but it is similar to spring feeding, since it could save your colony in the spring, even though it is fed just prior to winter. When the colony receives a good feeding in the fall, the syrup is, in part, consumed, since the nectar flows are over; but it is also stored for future needs. The feeding on the syrup now delays the consumption of the regular stores, thereby moving ahead the final moment when their capped honey will be exhausted in late winter or early spring.

In addition to adding medications to the two-to-one mixture of syrup, it is important to add one-half teaspoon of lemon juice to a gallon of syrup. The lemon juice helps retard granulation in the comb. The colony cannot use rock candy. The liquid gallon of syrup will add about 7 to 8 pounds to the winter stores if they consume it immediately. If you have a colony low in capped honey, three such gallons of syrup (twenty to twenty-five pounds) will do much to thwart starvation in the spring. The average colony must have at least 50 pounds after brood rearing ceases in the fall until early spring (middle February).

*The average colony needs 50 pounds from fall to spring*

In early winter, during the first ninety days following November 15, it is said that the average colony consumes about 10 to 15 pounds. The next thirty days, which is late winter, February 15 to March 15, they consume 15 to 20 pounds. So figuring 10 pounds for 90 days in early winter, 20 pounds in late winter, plus 20 pounds in early spring should tell you when they will run out of stores.

November 15 to February 15=90 days=10-15 lbs.
February 15 to March 15=30 days=20 lbs.
March 15 to April 15=30 days=20 lbs.
Total: 50-60 lbs. more or less

If you have finished feeding syrup to supplement their stores for winter (and you'll know because they'll stop taking it when they are full up), you are ready to continue "winterizing" by sprinkling granulated sugar over the inner cover until you can just about see the wood. The dry sugar will absorb whatever moisture forms on the under side of the outer cover, and it will be there, as emergency stores, when they break cluster on a mild day toward

*Sprinkle sugar on top of inner cover*

spring, should they need it. Place a good-sized rock on top of the outer cover to prevent winter winds from blowing it off.

*Check bottom two supers before winterizing for queen presence and disease*

Earlier in the fall, you will take certain steps prior to the formal wintering routine. Naturally, any surplus honey will be removed and processed by extracting, or stored in a warm room until you are ready to uncap, extract, and bottle. Then comes an inspection, which must be as thorough as possible. Get some idea of just how much honey there is between the two brood-food chambers. You can figure conservatively that a deep frame completely filled and capped will weigh about five pounds. After estimating the total, you'll have some conception as to how much syrup will be required. As you become more experienced, you can heft the hive from the rear and know if it is a little light or just right.

This inspection must reveal the queen to you; for unless there is a good queen present going into winter, your colony will not survive. If this inspection is in mid-September, you should find some brood present, since the queen is providing the makings of the winter bees at this point. Check carefully for disease or any signs of its having been there.

*Precautionary medication*

Commercial bee men are strong for powder dusting the tops of the frames of the upper hive body with a medication made up of 3 pounds confectioners' sugar, 6.4-ounce package of TM25 Terramycin and 15 teaspoons of sodium sulfathiazole. This mixture must be used judiciously, as it can kill larvae in the cells. One tablespoon is carefully sprinkled on the top bars of the frames, but only around the

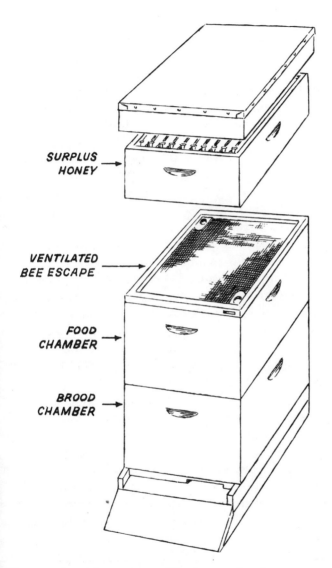

SURPLUS
HONEY →

VENTILATED
BEE ESCAPE →

FOOD
CHAMBER →

BROOD
CHAMBER →

The ventilated bee escape with the surplus honey super
above

outside and on the ends of the frames. This will prevent any powder from reaching the brood, generally found in the middle. This medication is again only a preventive, and not a cure, for both EFB and AFB. Use only one application in the fall and one in the spring. Make up only as much as you require, but in the proportions I have outlined.

*Add — Insulite_ board above_ inner cover_*

Insulite board (Celutex) is a moisture gathering and releasing medium. It will serve that function if a piece the size of the inner cover is placed between the inner and outer covers. It is an extra measure of insurance to keep the colony dry through the winter.

I have never advocated wrapping the colony with heavy, elaborate packaging. It appeared to be more of a deterrent than an assist to the bees. It is true that, when the temperature drops, the inside temperature will go down more slowly; but down it will go, until it matches the outdoor readings. When it warms up outside, the hive will respond too slowly for the bees to take advantage of the warmth to change their positions; or to reorganize honey stores to a more convenient location.

*Wrap hive in black tarpaper and add a windbreak*

I believe, however, that a single wrapping of black tarpaper will help the colony receive warmth generated by full sunlight in winter. Just be certain to push a hole through the paper to maintain the upper entrance hole. Wrap once around the hive with an optional single piece for the top of the outer cover.

The final step in preparing the colony for winter is the provision of a good windbreak against prevailing winter storms. Bales of hay, brush piled high, a tree line, or a burlap winter fence will protect the colony from sustained heavy winds. It really helps lessen the stress placed on the bees during the winter.

# 14

## Profits in Beekeeping

It is true that honey is the bottom line in beekeeping. Certainly a supply of your own honey for the family and friends is satisfying. Also, you might have a surplus which can easily be sold, thereby recouping a good portion, if not all, of your investment. That is only the beginning, since the bees will also be responsible for a large increase in the yield from your vegetable garden.

*Honey is the bottom line*

As the bees perform their daily task of gathering nectar and pollen for the colony, they pollinate many different fruits and vegetables. Ornamental flower gardens also benefit from the bees. Pollination does occur to some extent from other means, but complete pollination by our honey bees provides an abundance at harvest time everywhere.

*Increased pollination is a plus*

What about financial gain through beekeeping? Is there an opportunity for a person to develop a profitable business and still enjoy a "back to the earth" type of life? Investigation of all the facts would indicate a resounding affirmative answer.

*Honey can be profitable*

Honey is the first of several products having an excellent market potential available, and all are pro-

*Honey is replacing sugar in bakeries*

duced by the bees. Honey is rapidly replacing sugar as a sweetening agent, especially in commercial bakeries. It isn't that these mighty giants of the baking world are overly health conscious, but they have learned that honey absorbs moisture and can keep breads and cakes fresher over longer periods. Most small bakeries are anxious to have a steady supplier of this liquid gold for their requirements.

*Honey in demand with the individual consumer*

Honey is privately purchased by individuals from the beekeeper directly or most often through a good shop carrying natural foods. Independent grocers and gourmet shops alike are alert for the supplier of a good-quality honey attractively bottled and labeled. One need not be a salesman either to merchandise his honey to the stores. A personal appearance with a bottle of honey and an order book is all that is required. There is one more requisite, and that is a steady supply of the product once the sale is made. It is better to sell to fewer shops on a continuous basis than to have a stop-and-go supply for a large number of shops.

*Honey as a sideline to supplement your income*

The decision is the beekeeper's to make as to the scope of his operation. He can use his hobby as a sideline and make a considerable sum of money while continuing his regular line of work. Of course, it is certainly easier if that work is compatible to some extent with beekeeping. Teachers, professors, tree wardens and foresters, medical doctors, Wall Street brokers, lawyers, and corporation presidents, to name a few, are apparently in work that is compatible. They only have one thing in common: they are all sideliners in honey production and sales.

I know a beekeeper, a brilliant psychiatrist, who has eleven colonies. This chap markets about a thousand pounds of honey a season for just about as

Another production from the bees—beautiful beeswax candles

many dollars. Certainly he has his costs of operation. Colonies die over winters, equipment must be replaced or repaired, and the wax moths sometimes take their toll, so that new wax foundation is required.

*Some impressive success stories*

Then I have a friend who is a great bee master and is far more knowledgeable than most; he keeps about 150 colonies of bees, carefully attended. Though most old hands at beekeeping won't believe me, he has harvested almost ten tons of honey from his hives in a good year. His original investment was naturally quite large, and operating expenses are commensurate with an operation of that size. He also is a sideliner, since he is a craftsman and earns his main income from his craft.

Finally there are those members of the beekeeping

*Two people can manage 400 hives*

fraternity who are first-class honey producers. These people earn their income solely from beekeeping and run apiaries having a minimum of some four hundred colonies. Generally this size business can be managed by husband-and-wife teams; or frequently they are father-and-son operations. Larger commercial ventures can range into thousands of colonies, requiring many employees. One such business has 2,500 colonies spread over two states and employs about six people most of the time.

There is also a chap who runs about 1,500 colonies, primarily as a one-man operation except during intense honey flows. During bad years he only gets about 40 pounds of honey per hive; but the better seasons yield up to 250 pounds a hive. His apiary location is excellent.

*Location and logistics are crucial for the large operators*

Some of the really large honey producers have 10,000 or more colonies of bees, spread out in 200 or 300 bee yards over several states. The administrators of these businesses are experts in logistics and superb coordinators of vehicle movements and men. The harvest of honey from large apiaries of this type is put up in drums and tank trucks, which are delivered to the brand-name wholesalers. Such honey generally finds its way into the supermarket chains. The honey as a rule is not inferior in quality though it travels a much longer route from producers to consumers.

As you can surmise, honey production covers a wide range of business magnitude. You have heard of the cattle barons and the oil barons; wanna be a bee baron? It's up to you!

Remember that honey is but one of the wonderful products that the bees give us. Beeswax is another.

The white beeswax capping that covers the cells containing honey must be cut away by the beekeeper in order to expose the honey for extraction. This beeswax is accumulated and then melted down into convenient cakes, sized for storage or shipment. Tremendous quantities of beeswax are used by the cosmetics industry. Lipsticks, creams, depilatories are but a few of the products requiring its use. High-grade furniture waxes and floor polishes also incorporate beeswax in their formulations. Large quantities are used in the arts. Sculptors, painters, and batik workers seek out supplies of beeswax. The list is quite long, and you can see that the demand for it outstrips the supply. The beekeeper who produces honey will also have beeswax available after the harvest. On a pound-for-pound basis, it is more valuable than honey.

*Bees also produce beeswax for many uses*

The beekeeper need not seek out a market for his raw beeswax; it will find him. If he is industrious, however, he will learn to mold or dip his own beeswax candles, finding a ready market among friends or shops nearby.

*Making your own beeswax candles*

Pollen can be a third product useful in the aggregate of revenues for the beekeeper. I said "can," since the beekeeper must trap the pollen from the bees' legs as they enter the hive. The pollen is then contained in a small clean drawer beneath the trap until a sufficient amount is collected, or two days have elapsed, whichever comes first. If left on the hive for longer periods, the trap will cause retarded growth of the colony. A considerable amount can be collected in that short period, and if more pollen is wanted, the trap is moved to another colony. This avoids a hardship for any one colony.

*Pollen as another product*

Pollen is a self-contained protein food, easily digested, totally nutritious. Entire books have been written on the subject, and on pollen's beneficial effects upon human beings. A single pellet on the bee's hind leg weighs about 10 milligrams and contains roughly 2 million grains of pollen.* Each grain of pollen has the capability of fertilizing a fruit or vegetable, since it is the sperm of the flowers. I believe it is obvious that pollen can provide a gain for each of us. It is a relatively new field of endeavor in which marketing will play an important part.

*Propolis, another product*

Propolis, another by-product of beekeeping, is a material which the bees make and use as a glue. Hairline cracks in the hive and openings of any sort smaller than a bee space are filled with propolis by the bees. It appears to be a combination of resins from trees and other vegetation. Several countries outside of the United States have started work on propolis extracts, which are claimed to have antibiotic properties. The use is new in this generation but it has been mentioned in literature as far back as a century ago.

*Propolis as a medication*

I can personally attest to some of propolis's benefits, since it has become ritual, when I or my wife feels a sore throat is imminent, to chew on a small piece of propolis. While we are chewing the propolis, the saliva carries juice or extract down our throats. The ailment disappears. I have used propolis for easing headaches that accompany a sinus problem. One of our family members had an earache develop and again propolis was called upon successfully.

There is a company in the United States that buys

---

*Felix Murat, *Bee Pollen, the Miracle Food*, F. Murat: Miami, 1977.

propolis by the pound from beekeepers. They probably export it to the markets I mentioned earlier. At this time it commands a price twice that of beeswax, and beeswax is currently at an all-time high.

Royal jelly is produced by the bees to feed their brood and, of course, the queen. The queen receives royal jelly when she is first an egg, and she continues to be fed royal jelly throughout her entire life. Worker and drone larvae are fed royal jelly for only the early period of their larval state. *Royal jelly as another product*

Royal jelly is surrounded by a mystique, but its properties are well known today, thanks to scientists who have investigated its composition. Though it has been found to contain several vitamins, including B, C, and D, its antibiotic qualities have especially interested the scientific community. Further research is necessary before it will be lifted away from the fictional claims for its value and into an area of sound facts. The absence of these facts has not hurt the current market, and it appears that the market will grow as more scientific testimony comes to the surface.

Recently, medical circles have been making small noises which must be considered as interest in bee venom therapy as a treatment for rheumatoid arthritis. Several papers have been given on this subject, revealing, in part, that certain chemicals contained in the venom trigger human organic functions which produce materials, among them cortisone, which are helpful in neutralizing the effects of the disease. Here again, research is required before definitive results will establish the facts.

In addition to this therapy, whole bee venom is being administered to those individuals who are

*Experiments with bee venom*

allergic to bee stings. They are desensitized as a result of receiving a gradual increase in bee venom injections or stings. Future requirements for bee venom may call for beekeepers to supply it, after establishing methods of collection.

Any individual who pursues the art of queen breeding will find it extremely lucrative if conducted in a businesslike manner. Dependable sources of queen bees are constantly searched out by all who keep bees, whether for business or pleasure.

I'm sure there are many other ways of allowing this honey bee of ours to earn us a living. For example, a good beeman is always in demand among honey producers as are good photographs depicting the honey bee at work. The list goes on and on, and you will probably find other avenues open, which may not have been discovered as yet.

# 15

## *Winter*

On the upper half of this globe that we call our world, winter comes on December 21. It is the shortest day of the year for us, according to all the scientists; for you and me, it means the sun, which is furthest away on that day, now starts what seems to be a long trek back to us. Each day is just a little bit longer than the previous one, until it stares down on us at the beginning of summer, June 21. That is when it turns around and starts for the southern hemisphere again.

We all know that it is warmest in summer and cool to cold in the winter; but isn't it odd to find out that we go through the roughest part of the winter as days are getting longer and more sun should be warming us up? The reverse happens in the summer, because after Ol' Sol pours on the heat on June 21 he leaves that warmth behind, all the while seeming to travel away. The earth retains warmth as its great mass cools down slowly until much of the surface heat is lost.

As it cools, all nature takes the clue from Ol' Sol and, sensing the advent of fall, prepares for a period of rest. Our honey toilers twirl their antennae and receive the signal too! Preparation becomes more urgent, and about the end of November, and perhaps a little before, her royal majesty settles into the warmth of her bees' cluster. No more egg laying until about the middle of January. It may seem odd that the queen perks up and starts the year anew just one month after the sun has made the days a half hour longer, but that's what she does.

Once the queen begins her egg laying, it increases slowly each day; and each day requires more food for the brood and to maintain the temperature needed for incubation. Up to this time, the colony was using ten or fifteen pounds of honey, but now their consumption of stores will increase, so that from mid-February the bees will be using up two and even three times the twelve-pound amount every thirty days. Hopefully your bees have adequate stores of honey or sugar syrup fed at the end of summer. But is there enough pollen, which is so essential to brood rearing? I don't bother to find out and assume there is not.

During the first break in temperature after the middle of February, I supply my colonies with a pollen patty about six inches long and three inches wide. They are easy to feed and will insure your bees ample pollen and sufficient carbohydrates. They will raise a strong colony by the time the nectar flow begins.

Pollen substitute and pure pollen can both be purchased. The substitute usually consists of soybean flour, dried skim milk and brewer's yeast. It is available in powder form and is premixed. When pure pollen powder is added to the substitute, it is

*The queen stops laying from mid-November to mid-January*

*In early February the bees need more food*

*Add pollen patties after mid-February*

*Or add a mixture of pollen powder and pollen substitute*

much more appetizing to the bees, though not necessarily more nutritious. Add ½ pound of pollen to 1½ pounds of pollen substitute. Blend the powders well before you add 21 ounces of water. Work this mixture until you think your hands will work no more. Don't add any more water, though you will be tempted. The mixture will be finished when there is no more loose powder and it looks like caulking compound. Now you can add 2¾ pounds of granulated white (ugh) sugar. As it mixes in, you will find the entire mix smoothing out. What is happening is the liquefication of the sugar. It has absorbed water in the mix and as it (the sugar) liquefies it wets down the rest of the conglomeration. Mix it and knead it until nice and smooth. No lumps and no loose dry powder.

Sandwich the whole works between *two* sheets of wax paper, about 36 to 40 inches long. Roll out the mixture between the wax paper until you have one huge patty about three feet long and as wide as the paper. It should be around ¼ inch or less in thickness. If you place enough thicknesses of newspaper under it, it becomes easy to slice up with a sharp knife. I generally take a yardstick and lay it lengthwise down the center of the wax paper. I run a knife down the entire length, cutting it into two long strips each about six inches wide. (Our wax paper is 12 inches wide.) Then I cut across these strips every three or four inches. Take whatever you need and put the rest in the refrigerator, not the freezer. It is clean and it won't contaminate any foods in the refrigerator. Well sealed, it won't dry out.

Poke a half-dozen holes into *one sheet* of the wax paper, which will become the bottom side of the

*How to make a patty*

*Insert the patty in the hive*

patty. After uncovering the hive, smoke the bees down so none are on the top bars. Place the patty on the top bars, but be sure it is directly over the cluster. Now be sure to reverse the inner cover, ledge side down, to make room for the patty. Close the hive.

You have now started your second season of beekeeping, and you may be confident your bees will be powerful for the summer ahead. Every seven to ten days you will diligently open the hive and repeat the procedure, each time adding a new patty, even if some of the previous one is left.

*Preparing and repairing your equipment*

During your long winter, which becomes terribly short right after January 1 for every beekeeper, you will probably prepare any new equipment for the coming season. Learn to have frames and supers all constructed and painted before the active bee season begins. Don't hold a swarm just captured in one hand, and try to build a hive body with the other. It is also a good time to try construction of that gadget you saw in one of the bee magazines. You were too busy to make it up during the summer and you will be busy again. Reading about bees and beekeeping, as I hope you are doing now, is best done when the concentration is easiest. It can't be better than on a cold winter day in front of the fire.

*Getting to know your fellow beekeepers*

Take the opportunity also to look up your local bee club and join. The greatest exposure a beekeeper can receive is that of knowing and talking with another, whether he is a hobbyist or a commercial honey producer. Many new ideas and some old ones are expressed by different beekeepers, all having different conditions to work under. That makes it all the more interesting to you, since you can apply your own standards to the advice and accept, change, or refuse the various techniques offered.

Find out if there are any bee courses being sponsored locally. Sometimes a nature center will give a course that requires only your membership. Then too, the Y's and some adult-education programs are beginning to include beekeeping.

Large organizations devoted to beekeeping hold annual meetings of three or four days' duration in various parts of the country. You'll never know the fun you missed until you attend one of these get-togethers with the greatest of people—beekeepers. Eastern Apicultural Society is one of these organizations, and anyone can handle a four-dollar membership fee. The Western Apicultural Society is just forming at this writing, and I'm sure I heard rumblings about a "central" and a "southern" in the planning stages. There is also the great American Beekeeping Federation, which serves the beekeeping public well. They protect the purity of honey, marketed at large, by checking imported honey for additives, as well as keep track of everyday changes in governmental policy as it pertains to the honey industry. So get together with one of the many good groups of people and enjoy the company of other beekeepers.

*The advantages of bee organizations*

When the winter aconite and the crocus push aside some crumbled earth and show their perky petals amidst the snow, look to your colonies. Inside the hive impatience is popping. On the first mild day, though snow still covers the ground, the bees will be flying. They won't go far, but suddenly the snow will be dotted with all the colors of the retained feces. This is their cleansing flight, and the hives too receive their shower of color. Don't fret,

*For bees, spring comes early*

because after the first good rain they'll come clean again.

Any who have Bee Fever have nursed their bees from the package stage to a bustling beehive. Ask them and they will tell you, "I love my bees." If you find yourself in the company of a beekeeper after a fierce summer storm, you'll see him or her rush out to check on the bees' well-being. Ask him and he'll say, "I love my bees." Ask any beekeeper and that person is sure to tell you, just as I do, "I love my bees" . . . and so will you. As the queen of all bee-keepers once said to me, "Beekeepers are truly very special people!"

# Epilogue

As spring unhooked the talons of winter, and the snow melted on our roof's south slope, it gurgled its way through the downspout into the still half-frozen ground. And I was glad, for my first hive stood gleaming white like an empty lighthouse, silently waiting for its inhabitants. As soon as it was finished, I returned to the house on Elm Street where the bee lady lived, to learn more about beekeeping and, secretly, to learn more about Edna Erickson herself. Why couldn't I see her? Why was she hiding behind the curtains to her room?

My visits were usually on weekends, since I traveled on business during the week. My schedule was fairly busy, as I covered the northeast corner of the country and was a frequent customer of the airlines. At first I was not aware of how anxiously Edna awaited those visits. Her curiosity was almost insatiable. The details of each trip were unfolded as she plied me with pointed and exacting questions. She became interested and excited by descriptions of places considered commonplace by most people. Only then did I realize I had become her window on the world.

And so a pattern evolved; one week followed upon another, and as each Friday night found me returning home from my sales territory, weary of planes, trains, rental cars, and motels, I looked forward to "seeing" my queenly lady the next day. It was a strange kind of pairing: the traveling salesman and the bee lady. The curtain barrier to her bedroom gave our meetings a clandestine atmosphere, especially when her arm slithered into view with an offering of honey candy.

I usually arrived late in the afternoon; but one Saturday I found myself at Elm Street just after the noon hour. Doris, whom Edna referred to as her legs, brightened visibly as she opened the door, "I'm so glad you're early today," she proclaimed with evident pleasure. A moment later I understood her happy frame of mind. She escorted me to my now familiar chair just outside the curtains and explained pleasantly, "I need a few things at Brown's store, which I must pick up today before five o'clock; tomorrow, when I normally do my shopping, they are beginning their vacation. I won't be long," she continued cheerily, "but please take care of anyone who comes for supplies, if I'm not back" . . . and like a fairy, she was gone.

"Well, tell me where you've been this week, and how come you're early today?" Edna's resolute voice came through the curtains with great eagerness. I laughed to myself at the double-barreled set of disconnected questions.

"I'm not really sure why I'm here before my usual time," I said dreamily, almost to myself. Her question and my answer catapulted my brain into one of those swift, chainlike series of thoughts that speeds through your mind in seconds and takes minutes to relate. I realized I too was eager to see Edna; to perhaps catch the moment that curtain would accidentally open, or even to start my verbal campaign to have her open it. That curtain was only a slender barrier, but it did separate our beings physically. It would have to go, I resolved silently.

The next half hour I spent describing Buffalo's airport and its vulnerability to snowstorms. I had been there all right, an entire extra day, cooped up in a motel room, waiting for the airport's snowplows to get ahead of the snow on the runways. The air traffic was at a standstill, shut down, and nothing moved. When we finally did leave, the plane flared the snow away from it just as a great ship's bow cuts through the waves.

I imagined Edna's eyes widening as I related the story of that rather hairy departure from snowbound Buffalo; the New England area had only received a light dusting from this same storm. So I was surprised when Edna said efficiently, "The nectar flows will be good this year; for when we get as much snow as we've had all winter long and then have a good fall of snow this late in the winter, the ground has an overabundance of moisture."

The grinding front doorbell rasped out an announcement, which jolted me from mind pictures of water trickling through the soil, seeking the thinnest rootlets to penetrate.

Edna's visitor filled the doorway. His oversized red beard, which equaled his bulk, danced as he inquired politely, "Is this where one may procure the equipment to pursue the art of apiculture?" Flabbergasted by his question, couched in formal courtesy, which flowed from this rough and ready individual, I nodded and stuttered an affirmative answer.

Most of my regular business encounters were with officers of large corporations, and the transactions were frequently in six figures and better. I reflected on my own purchase of bee supplies from Edna via Doris several weeks earlier and realized I had spent about a hundred dollars to equip myself and provide a complete hive to house the coming bees. How shall I address myself to this potential customer of Edna's? I mused silently; and then said aloud, "Let's go out to the barn where the supplies are stored."

I led him through the house to the back door, which opened on a small patio, enclosed by the house and the barn in the rear. We entered the barn and my eyes were accosted with a disorderly array of boxes piled on the floor. It was my first visit to the stockroom, and I suddenly realized I had not thought very much about how Edna and Doris managed to manipulate these heavy boxes containing the lumber of beekeeping. "It's a little messy because a truck just arrived and we haven't stacked the delivery yet," I mumbled apologetically. Have to do something about this, I decided, as I selected the required equipment for the customer. "Will you accept my personal check?" he said gently, as I stacked the boxes in his burly arms. Recalling that Doris had taken my check in payment previously, I assured him it was okay. But did I have the right to approve of this man's credit and payment, I asked myself? After all, I really didn't know if the "girls" had a system of some kind to qualify their customers; and this one, Conrad Claxton, was surely an enigma, appearance of a lumberjack and the manners and diction of a fine gentlemen. Inside the house, I showed the components of the hive to Mr. Claxton. I followed the format Doris used when she explained to me how the equipment operated. Twenty or thirty minutes later, he wrote a check for the required amount. It was written on a Canadian bank. My heart sank. I couldn't accept it without Edna's approval! "I'll only be a moment," I announced efficiently, heading for Edna's bedroom.

"This man just paid me with a check drawn on a foreign bank. Will you accept it?" I blurted, and then gasped as I realized I had thrust the check under her eyes, which were locked with mine. I had unthinkingly burst into Edna's secret cubicle. Her violet eyes twinkled mischievously as she plucked the check from my frozen fingers. "Canada is between the same two oceans as we are, isn't it!" she stated impishly, with a humor that I learned later was limitless. "Sure, his check is all right," she continued. "I've been accepting beekeepers' checks for nearly fifty years and I have yet to receive a bad one. Beekeepers are very special people," she proclaimed with evident pleasure. Not a word was said about my intrusion into her private world. I was so relieved by her acceptance that I could have kissed her; and she looked kissable, even at seventy-five.

Edna's satin-smooth complexion was delicately shaded with hues of baby pink high on her cheekbones. She wore a Swedish crown, two coils high, made of carefully braided flaxen hair, neatly secured. "She is a beautiful woman," I told myself silently. Beautiful, yes, but her strong, straight jawline and jutting chin displayed determination and strength too. My eyes slid down the length of her form, outlined by the covering bed linen. I wondered about her legs, as my eyes rested upon the two little humps in the sheet made by her feet.

"They just don't work," she whispered softly, though not sadly. "I was seventeen when polio put me down," she explained hollowly. Suddenly she brightened and whispered victoriously, "But my spirit, never!"

She had been badly handicapped by this dread disease, but with apparent inner strength had overcome or, at least, moderated the effect of her disability. I suspected her strength was laced with faith. There were many questions pushing to be asked, but I decided to let them sleep till later in our relationship.

I helped Mr. Claxton, our new beekeeper, to his car, with all the boxes, and assured him that he could return at any time to ask for help and advice on all matters connected with the husbandry of honey bees.

After he left, I found myself in the barn, stacking boxes one above the other, each stack displaying the same item number. The enterprise expended more physical effort than I had anticipated, and each time I rested to catch my breath I mentally explored Edna's situation. One severely immobile elderly lady, assisted by a very mobile elderly lady, sustained herself in this little old house with an income derived from the hobby of beekeeping. If that feat, in itself, was not enough, how did they handle the truck deliveries received from various equipment manufacturers? How were inventories and stock levels maintained with that scrambled merchandise in the barn? My interest in Edna grew stronger.

Eventually the inventory of boxes was segregated into orderly piles. Even before I returned to the house, I succeeded in making a rough count of the merchandise. I revealed a shortage of a couple of items and it gave me a great excuse to reenter Edna's room.

I passed my hand through the curtained doorway and called her name softly. She grasped my hand firmly and drew me into the room. I handed her my roughed-out inventory sheet and waited. After a few moments, she looked up and said uneasily, "We don't have any head veils according to your list." "That's right," I answered. "We also don't have any smokers or gloves." Edna reached for a catalog and formed a purchase order. I watched and learned.

At that moment, I think an unofficial partnership began. I could only help on weekends, and of course it was truly a labor of love. I took nothing from Edna but her wisdom, which she willingly shared, along with her marvelous humor and unbelievable courage. She also gave me her affection and a new meaning to the word "strength." Our relationship deepened, and during the years that followed we spent many beautiful and private hours together.

I learned that Edna's father had been a beekeeper. They had lived in this same house—in fact she had been born in it, an only child—and she had become his full-time assistant, getting about on crutches. Her main duty was clipping the queens' wings, for her touch was delicate and his hands were rough. After her father died, Edna too kept bees, for the city had not yet crept around the house. Eventually her crutches no longer were enough to help her navigate, and she had had to give up keeping bees. But the regional dealership held by her father had continued with her since his death many years before.

Edna's true assistant, Doris, her mobility, her "legs," was struck down by a fatal heart attack a couple of years later; and Edna was suddenly alone and very frightened. I played hooky from my business frequently, to be sure her little business would not founder. Edna depended upon the income it provided.

One morning around 6 A.M., she woke me with an urgent phone call. It was the first time I felt her lose her stride. "I don't know what I'm going to do," she breathed frantically into the phone. I tried to reassure her but it wasn't working. I decided to see her when she explained that her friend, Ernestine, could not stay with her for the entire day but only till noon. Many years before, Ernestine had been one of the many people that Edna had helped through a desperate period.

Edna and my wife, Anita, had become very close. I asked Nita to come along, for I felt that if I had to leave on any errand she could cover the remainder of the day. Little Doris was really missed.

We shifted the "watch" with Ernestine, and while leaving she suggested that she knew of a replacement for Doris. A friend, who had been a companion to an elderly person, found herself out of work when that person passed away.

She proved to be quite suitable as a helper to Edna and as a replacement for the previous Doris—especially since her name turned out to be Doris too! The new Doris was younger and stronger. Edna needed someone like that, since she required help in order to change positions, as well as to be removed from the bed when linens were changed. It was a great excuse to get her into a wheelchair and out into the parlor. Edna had come out of hiding at last and now seemed to enjoy her life more than before. She talked with her clientele in the parlor as she sat in her wheelchair for fairly long periods. Her spirits seemed to soar. Edna was getting younger and even more vibrant. It was a beautiful thing to see; and when she had a visitor, a potential beekeeper, her enthusiasm was contagious as she spoke eloquently about the bees.

After a couple of weeks of coaching, the new Doris began caring for the needs of the "store." She helped the customers, though she relied heavily upon Edna's help. It was working though and Edna seemed happy.

I had begun to wonder about me, however, since I now was spending a couple of days of each week with Edna and Doris, not counting Sundays. I had about a dozen and a half hives of bees by this time and was producing a thousand to two thousand pounds of honey a season. It was great fun giving away little half-pound jars to my friends and selling the rest whenever the mood struck me. The hobby was certainly self-liquidating by now, and all of my investment had been returned. It set me to thinking more seriously about the commercial aspects of beekeeping. Could I earn a modest living from the bees? Maybe, just maybe, I could work out a way to lessen the strife and stress that surrounded me! As with many other good and constructive thoughts that we all have, I lost the motion of it when pursued by the need of the moment; the absolute necessities of life crowd out the occasional gems that light up in our heads. But the thought of adding more and more hives to my holdings kept returning dreamily to my mind for the future. The question of how many I would need could be answered later when it really came to pass. After all, I reasoned to myself, you'll just keep adding hives until you have enough bees to make enough honey to make enough money! It couldn't be simpler than that, could it?

I think it was late on a Friday afternoon when Doris called me, and when she spoke my name, the catch in her voice sent an electric shiver through me. "What's wrong, Doris?" I feared the worst. "Edna," she started and paused and then gasped, "broke her hip and I've already called for an ambulance." I told her I would be at the hospital as quickly as possible.

The next few weeks were something of a nightmare, though again Edna's humor triumphed. I recall that after she was x-rayed in Emergency she was placed on a rolling stretcher for transport to her room. A strapping blond giant of an orderly provided the motor power for the stretcher. He pushed it from the head end so he was looking straight down at Edna's face. Apparently he had a date, and since the hour was, by now, quite late, the hospital halls were empty. He shifted into third gear and was literally trotting as he pushed the stretcher, and I was trotting alongside. Edna craned her neck a little to see him better and then quipped briskly, "Say, young feller, what be the speed limit around here?" We all broke up. Here she was in the hospital, facing a rough operation, certainly in some pain, but still maintaining a sense of humor.

A week after Edna's hip had been pinned together, she didn't seem to be making any progress. She was much weaker and her face had lost its beautiful luster. We were assured this was normal and we could expect improvement as soon as the shock of the operation wore off.

Ten days later, Edna was moved into a beautiful convalescent home, high on a hill. The shock of the operation had not yet worn off and Edna seemed to be slipping away from me. She recognized me, when I came to visit each day, and even looked for her hug and kiss; but it wasn't the same. She was smaller, frailer, and the outline of her body under the bed linen seemed vague.

When I visited Edna at the convalescent home, I customarily peeked around the corner as I reached her room, in case she was asleep. It was a golden August afternoon and the window curtains made lazy changing patterns on the ceiling. Edna's eyes were bright violet and focused above her. One arm was on top of the covers, the elbow bent, with her index finger pointing, almost waving at the ceiling shadows. Her lips were moving and I inched my way closer. A great empty feeling started in my stomach, dropped into my groin, and made my legs limp as I heard her whisper, "I know, I know, I'm ready." Then she turned her head toward me as though she knew I was there all the time. Edna beamed a smile at me and said in a clear cheerful voice, "Ed, I'm so glad you came today. Come closer." Her hand reached out for me and I held it. "I have had a presentiment today and I did want to see you again," she continued. "That's nonsense, Edna, you look better than you have in the last couple of weeks. You're on the mend now," I assured her. It was true; she looked well, even seemed to have some color.

Her grip on my hand grew tighter and she pulled me down. Her eyes never looked away. I unhooked the side bars on the bed to get closer. She kissed me and whispered for me to leave. I held her head in both hands, smoothed her hair, now loose, and then I kissed Edna. As I looked at her hard, I knew I didn't have to say it. She knew I loved her. Without a single word, our eyes said goodbye; and I left. Edna died before dark, that same day.

After Edna was laid to rest, I reinvolved myself in the maelstrom of my regular business. It was back to credit-card lunches in five different cities in just as many days. It wasn't easy even to try to stop thinking about Edna. It turned out that I was never going to be able to do anything like that. A few weeks had passed when we were notified to appear for the reading of Edna's will.

In the company of about two dozen of Edna's oldest and closest friends, as well as some very distant relatives, we heard the reading of the first bequest. Edna had left to me her beekeeping-supply business: inventory, customers, and franchise with one of the major suppliers.

Her attorney made it clear that I had to move the inventory out of the house and barn promptly, since they were going to satisfy the will by selling everything, except the bee supplies. Where was I going to put a barnful of beekeeping equipment?

Arrangements were made to truck the merchandise to my home and store it all in our garage. When the truck unloaded the goods and left, Anita and Ed were the best-equipped hobby beekeepers for a thousand miles around.

As I looked around the garage it suddenly hit me! There it is! All the hives I could want, served up in scores of boxes. "More hives, more honey—more money." I was jolted from my bee-baron fantasy by a heavy fist on the garage door, "Anyone home?" thundered a deep voice. I raised the overhead door and a barrel-chested man in a plaid shirt stepped into the garage. "This must be the place," he declared with certainty, eyes roving over all the boxes. "Have to get some bee hives and they told me up at Elm Street you had 'em here!" he concluded. During the next few days, this kind of visit was repeated a number of times, by an assortment of people. Each seemed to require different kinds of bee supplies.

I saw the compass needle spinning round and round in my head; and then it slowed, coming to a stop 180 degrees from where it began. It pointed my life in an entirely new direction! Perhaps this was my chance really to change the pace of our lives; Anita and I would have to have a board of directors meeting and decide where we'd go from here.

The vote was unanimous. We took the plunge and changed our lives. Some of our closest friends thought we had gone balmy. "Bees! Are you crazy?" I heard it so many times that for a short while I too began to wonder if I had lost my judgment. I didn't lose it; instead, I gained hundreds of new friends. The bees have brought me together with so many wonderful people. Our little supply business has prospered at a slow, easy pace. We have been able to share our enjoyment with hundreds of beekeeper people, as well as hundreds of non-beekeepers. It is great to watch the faces of those unacquainted with bees as we relate many fascinating stories of bee behavior, and their inestimable value to the human community.

Besides teaching beekeeping courses, we are knee

deep in beehives and produce our own brand of honey, which is enjoyed by many. Nita makes beautiful beeswax candles, which are snapped up by all who see them, especially when we are invited to exhibit at a country fair. Our live observation beehive is a great attraction, educating the young and old alike. Graphic exhibits make our display a popular spot in the fair and Nita and I explode with the joy of sharing our beekeeping with so many. Finding such happiness in your middle years is almost more than you can handle. Just to say it is gratifying is an understatement.

Most of all, I think I enjoy the freedom. I can do anything that I want to do: build anything I want to; fix anything I want to; and I can be home when I want to! And, believe me, that's important to me. Living at home, not coming home just to sleep and leave, is a large freedom.

The most beautiful of all is our wondrous bees. They brought us to a healthy way of life. All of nature has come up close to us and we have become a part of it. How else can it be? Aren't our needs the same as those of every living creature? When I am in a bee yard and it is spring and warm, I watch the bees pour out of their hives, scrambling upward on their way to find their precious nectar, and I feel enriched because I am part of it.

The bees have brought us all of this and friends too. Friends who visit us for bee supplies and come laden with home-baked breads, jams, jellies, and baskets of home-grown fruit. And Anita shares with them a snip here and a snip there from her lovely gardens. It's a joy to live and share. Anyone can do it. You can too! Edna shared with me.

# Acknowledgments

I never could have completed this book without the unswerving faith and confidence that Anita, my wife, expressed each day. She was patient and supporting through those endless hours she spent at the typewriter transcribing my scrawls.

This book was inspired by Buz Wyeth, my editor, who generously gave me of his time and knowledge, without which I could not have shared my joy of beekeeping with those who read this little book.

Out of the hundreds of friends I have, who are beekeepers, one came forth, when he discovered I was writing this book, and asked if I had a title. I had not, and my fellow beekeeper, R. "Tek" Nickerson, suggested *The Queen and I*. I smiled when I heard it; it stuck, and I am eternally grateful to him for his sincere interest and thoughtfulness.

Another new-found friend, Lewis Robins, who is a brilliant author, inventor and thinker, immediately gave me the clues necessary to create even the smallest literary work once he discovered what I was about.

Walter Medwid, the director of the New Canaan

Nature Center, provided me with the platform to teach others about basic beekeeping. The many courses taught at the Nature Center did, in fact, become the body of the book.

A wealth of information contained in the book comes from the many fruitful hours spent with my Danish friend and master beekeeper, Anthon Pedersen. He has been my mentor these many years. His knowledge of beekeeping is boundless.

My gratitude is without limits for my many friends and their contributions: Herbert Tanzer for his clairvoyant insights and his "compassion." Howard Stabin, who, like the impresario he is, stimulated me to both deeper and greater expression of my innermost feelings. And wonderful Jim Baker, who, as one of my early students, intoned, "You really ought to write a book!" Thank you, Jim, I did.

I know that I am only one of countless numbers who have been moved by the deep love and strong philosophies expressed by Richard Taylor in all of his writings. He is truly inspirational when he talks to nature.

The warm and ebullient gentleman, Calvin Diehl, who illustrated this book, proved to be an expert in his craft and sharing in his patience with me each time I described what was in my mind's eye.

"Professor Al" Avitable provided the needful eye to the science and the gentle humor involved with keeping bees. This gentle man of science keeps his spirit light, even when he is deeply drawn into the cell.

And lastly, two people who brought me closer to the spiritual workings of nature through their sincerity and love. Eugene Graf, whose life is dominated by his love for all that lives, and Chris Klotz, who left this world for another but who first, in his nineteen years, touched and left the warmth of sincerity with everyone he knew. His earnest love for all of nature added greatly to my own.

# Index